INTRODUCTION TO MEDICAL CHEMISTRY

D. MacLean D. Evans
UNIVERSITY HOSPITAL OF WALES
CARDIFF, WALES

John Bowen Jones
LOUGHBOROUGH COLLEGE OF EDUCATION
LEICESTERSHIRE, ENGLAND

HARPER & ROW, PUBLISHERS
New York Hagerstown San Francisco London

Sponsoring Editor: Joe Ingram
Project Editor: Lois Lombardo
Designer: Rita Naughton
Production Supervisor: Will C. Jomarrón
Compositor: Progressive Typographers, Inc.
Printer: The Murray Printing Company
Binder: Halliday Lithograph Corporation
Art Studio: Danmark & Michaels, Inc.

Library of Congress Cataloging in Publication Data
Evans, David MacLean Demetrius.
Introduction to medical chemistry.

Includes index.
1. Biological chemistry. 2. Chemistry, Medical and pharmaceutical. I. Bowen Jones, John, joint author. II. Title.
QP514.2.E9 612′.015 75-30629
ISBN 0-06-041921-0

To Betty and Mari

Contents

Preface

Chemical reactions underlie all living processes. Recently molecular biologists have been unraveling the code that enables molecules to transmit genetic information from one generation to the next. This is an example of very complex biochemistry. But even a rudimentary understanding of chemistry can provide an insight into many living processes.

The opening chapters of this book provide a summary of basic chemical concepts; previous knowledge of chemistry is not essential. The chemical elements from which life is formed are then considered, with particular reference to their biological context. This is followed by a discussion of the organic substances, once believed to be produced only by living organisms but now known to obey the same laws as inorganic substances. From these we proceed to examine the way these substances are organized into biochemical pathways to provide the coherent pattern of living processes. The study of body fluids comes next; their very nature provides a kind of review of these processes. The book ends with two chapters on radiation chemistry and its growing importance in the medical field.

We wish to thank Janet Brown, G. Haddock, and R. Toogood for the photographic illustrations; Yvonne Baldwin for her excellent line drawings on which many of the illustrations are based; our biochemist, P. Henry for his unstinted help; Dr. C. A. Joslyn for his suggestions on applied radioactivity; and especially Dorothy Symons for her untiring assistance with the manuscript.

D. MacLean D. Evans
John Bowen Jones

1

Historical Introduction

Chemistry evolved out of man's need to make weapons, woman's desire for ornaments and their joint need of vessels, utensils, and embalming fluids. By 2000 B.C. the ancient Egyptians and Sumerians had already discovered how to extract gold, copper, silver, lead, and iron from deposits. They could also fix dyes to clothing materials with mordants of metallic salts and make coats of many colors.

By the beginning of the Christian era chemistry had become a recognized branch of learning in the city of Alexandria where it was known as the divine art. An early name for Egypt was Chemia, and when the Arabs invaded Egypt they adopted its divine art as 'al chemia' or alchemy.

The alchemists of the middle ages were mainly occupied in attempting to transmute base materials into gold. Paracelsus (1493–1541), a rebel physician and alchemist, hotly criticized medical practice of his day and founded a sect of medical chemists called Iatrochemists, who applied chemistry to the service of medicine. In particular, they tried to prepare the elixir of life, which would cure all diseases and also confer perpetual youth. Although unsuccessful in their main endeavors, they did incidentally discover many new substances such as alcohol, mineral acids and metallic salts.

Van Helmont (1577–1644) represents the transition from medieval alchemy to modern chemistry. Born of noble origin in Brussels, he qualified in medicine, married and then spent his life in careful chemical experimentation. He was the first to recognize clearly the production of gas in chemical processes. He derived the name gas from the Greek word "chaos" to represent the wild motion of its particles. He discovered that when he corked up limestone and acid in a bottle to produce carbon dioxide, it burst the bottle. So he called it gas sylvester—the wild untameable gas of the woods. He also concluded, wrongly, that a gas could not be contained in a vessel.

Robert Boyle (1627–1691) has been called the father of modern chemistry. He set up a laboratory at the back of his sister's house in Pall Mall, London, where he developed chemistry as a science in its own right and not just as a means of making gold or medicines. His chief contribution was the introduction of the scientific method, every theory being put to the test of carefully planned experiment.

Boyle developed the concept of chemical elements. He defined an element as a substance which cannot be split by any chemical means into simpler substances. The Greeks had originated the idea of elements almost 2000 years earlier, but believed them to be earth, water, air and fire. None of these is an element in the chemical sense. They can, however, be looked upon as a poetic representation of the three states of matter: solid (earth), liquid (water), and gas (air). Temperature (fire) is a major controling factor; heat converts ice (solid) into water (liquid) and then into steam (gas).

Once the nature of chemical elements had been understood, the stage was set for the rapid advances in chemistry which took place during the next 300 years.

2

What Is Chemistry?

Chemistry is the study of substances, their properties and structures, and of the reactions by which they can be changed into other substances. By *structure,* we mean the arrangement of the minute particles of which a substance is composed.

MATTER IS TINY PARTICLES

Why do we believe that matter consists of tiny particles? We are all familiar with an important property common to all gases and vapors, the ability to spread or diffuse. Kitchen smells, unless we are careful, quickly pervade the whole house. The odor of disinfectant can be detected some distance from the place where it is used. If we think about it, the best explanation for this is that gases and vapors consist of tiny invisible particles that are constantly moving about in all directions.

Support for this idea is also given by the diffusion of liquids. If, for example, some bromine (a very dark red liquid) is placed under some water in a bottle we notice that after a day or two the bromine has spread evenly throughout the water giving it a uniform red color. We can explain this quite easily if we assume that the bromine and the water are made up of tiny particles which are continually moving about. Diffusion between solids is far less common than between liquids and gases. As a rule, the particles of a solid do not move about in all directions; they simply vibrate.

This theory of moving particles, which is supported by a great deal of other evidence, has proved extremely useful in explaining many properties of substances. It is one of the great theories of science and is known as the kinetic theory (Gr. *kineo*, I move).

TYPES OF PARTICLES

How many kinds of particles are there? Broadly speaking there are only three: atoms, molecules and ions. Before describing them we must first consider what is meant by elements.

Elements

These are different from all other substances in one very important respect: they cannot, apart from the radioactive elements, be broken down into simpler substances. Water is not an element because if we pass an electric current through it we split it up into hydrogen and oxygen. Similarly, common salt (sodium chloride) is decomposed by electricity into sodium and chlorine. Water and sodium chloride are called *compounds*.

There are at present 104 known elements; 89 of them occur naturally and the rest are man-made. A list of them is given in the Appendix.

Atoms

These are simply the smallest particles of an element. The Greek philosopher, Democritus, over 2000 years ago, originated the word atom, meaning a particle which could not be cut into any smaller pieces (Gr. *a*, not, +Gr. *tomos*, cut), but it was John Dalton from Lancashire, England, who first used the idea of atoms to explain chemical phenomena. His atomic theory, which he produced in 1807, is the cornerstone of modern chemistry.

Since Dalton's time, physicists have discovered that atoms are themselves built up of tiny particles of over thirty different kinds. However, only three need concern us. They are the proton, the neutron and the electron. We shall deal with them in Chapter 4.

Molecules

The atoms in many substances join together to form groups which we call *molecules*. Some molecules contain only two atoms; others, known as giant molecules or macromolecules, contain thousands of atoms. Chemists are able to determine by experiment and calculation not only the *number* but also the *arrangement* in space of the atoms in a molecule, in other words the shape of a molecule. The shapes of a few molecules are illustrated by the models pictured in Fig. 2.1.

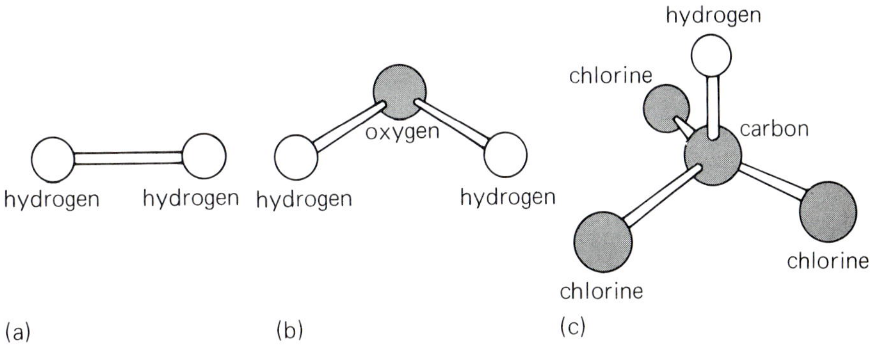

Figure 2.1
Models of molecules (a) hydrogen H_2; (b) water H_2O; (c) chloroform $CHCl_3$.

The rods joining the spheres represent the *chemical bonds* which hold the atoms together. We shall find out later what a chemical bond actually is. The properties of a substance are largely determined by the bonds between its atoms and by the way the atoms are arranged.

Representation of atoms and molecules

Undoubtedly, the best way of representing molecules is by the use of models or pictures of models. However, we often do not need to show the shape of a molecule, only the numbers and types of atoms it contains. We can do this quite easily using the molecular formula of the substance. H_2O is the molecular formula of water. The symbol H represents one atom of hydrogen and the symbol O one atom of oxygen. The number in a formula always refers to the symbol preceding it; the formula H_2O tells us that a molecule of water consists of 2 hydrogen atoms and 1 oxygen atom. The molecular formula of oxygen is O_2. This tells us that a molecule of oxygen contains 2 oxygen atoms. What does the molecular formula of glucose, $C_6H_{12}O_6$ tell you about the glucose molecule?

Ions

Atoms or groups of atoms which carry an electrical charge are called *ions*. Their nature will be described in the next chapter.

3

Electrical Attractions and Repulsions

EXPERIMENT 1

Rub a plastic pen with a piece of cloth; your sleeve will do nicely. You will find that the pen will attract certain materials, such as small bits of paper, a cotton thread, your hair, and more spectacularly, a fine jet of tap water. We do not know what causes these attractions, but this does not prevent us from giving them a name. We call them *electrical attractions* to distinguish them from magnetic and gravitational attractions which we are also unable to explain.

Now wipe the rubbed pen with your hand and see if it will still pick up bits of paper, etc. You will find that it will not.

When the pen is capable of electrical attractions we say that it is *electrically charged* or that it carries *electrical charges*. Wiping it with your hand *discharges* it.

EXPERIMENT 2

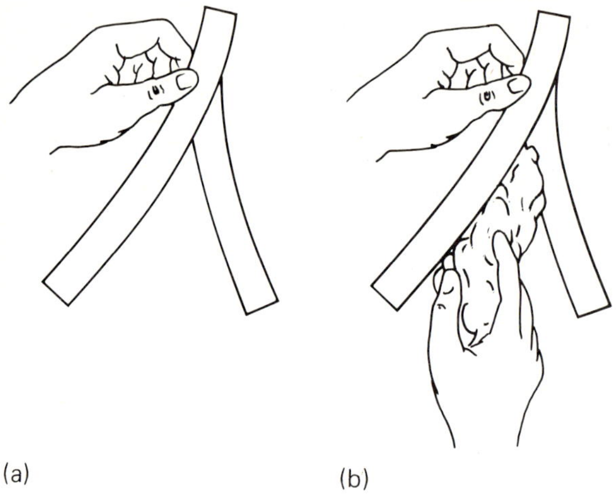

Figure 3.1
(a) Charged strips held together at one end; (b) handkerchief placed between two strips.

Cut two strips of polyethylene, each about 12 inches × 1 inch, from a polyethylene bag or sheet. Place them on a flat surface and electrically charge them by stroking them a few times with your handkerchief. Check that the polyethylene is charged by seeing if it will attract a small piece of paper.

Now hold the charged strips together at one end (see Fig. 3.1(a)) with the "stroked" surfaces facing each other. You will observe that the strips repel one another.

Still holding the two strips together, place the handkerchief between them as shown in Fig. 3.1(b). You will notice that the polythene is attracted to the handkerchief.

Now slowly withdraw the handkerchief starting from the junction of the strips, so that it gently brushes their inside surfaces. You will find that the strips no longer repel each other; they are no longer charged (more than one brushing may be necessary to destroy the repulsion completely).

As a result of Experiments 1 and 2, we can make the following deductions.

1. The charge on the handkerchief canceled out the charge on the polyethylene strips. There must therefore be two kinds of electrical charge which are "opposite." We can denote opposites by using the symbols + and −. We say that the charge on the rubbed polyethylene is negative, and that on the rubbed handkerchief is positive.

2. The two strips of polyethylene both carried a charge and they repelled each other. Therefore, negative charges repel one another. It can also be shown by experiment that positive charges repel one another. In general, we deduce that *like charges repel.*
3. The negatively charged polyethylene was attracted to the positively charged handkerchief. Therefore, *opposite charges attract.*

What has all this to do with chemistry? Quite a lot! You will remember in Chapter 2 we said that some substances were built up of particles called ions. We are now in a position to understand what an ion is. It is simply an electrically charged atom or group of atoms. Sodium chloride consists of positive sodium ions and negative chlorine ions. We can see right away what holds the sodium and chlorine ions together in sodium chloride, it is the mutual attraction of a positive charge and a negative one. This attraction is called an ionic bond. We shall see in Chapter 5 how particles acquire electrical charges.

Reference was also made in the last chapter to the three subatomic particles, the proton, the electron and the neutron, from which atoms are built. The proton is positively charged, the electron is negatively charged and the neutron is, as its name suggests, electrically neutral, that is, uncharged. The building of atoms from these three fundamental particles is the subject of the next chapter.

4

Structure of Atoms

In Chapter 2 we said that atoms were built up from smaller particles. Many of these are unstable but there are three stable particles which are fundamental in atomic structure. They are the proton, the neutron and the electron

	SYMBOL	RELATIVE MASS	ELECTRICAL CHARGE
proton	p	1	+1
neutron	n	1	0
electron	e	1/1840	−1

ATOMIC STRUCTURE

An atom has a tiny central nucleus which contains the protons and neutrons. The term nucleon refers to a nuclear particle, whether it is a proton or a neutron. The rest of the atom is thinly populated by electrons which revolve around the nucleus. The number of electrons is equal to

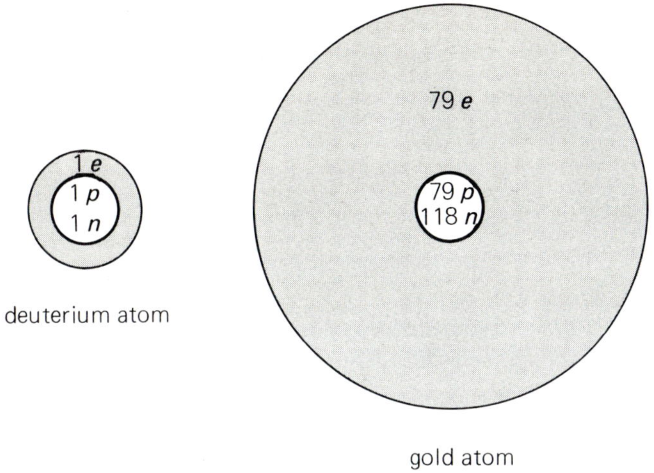

Figure 4.1
Deuterium atom and gold atom. Representation of particle structure: e, electron; *p*, proton; *n*, neutron.

the number of protons so that the atom as a whole is electrically neutral. Atoms of deuterium and gold are shown diagrammatically in Fig. 4.1.

Because the mass of an electron is only about one two-thousandth of the mass of a nucleon, almost all the mass of an atom is concentrated in its nucleus.

Let us summarize: An atom consists of a relatively tiny and heavy, positively charged nucleus around which negatively charged electrons orbit.

Atomic number

This is the number of protons in the nucleus of an element and it is the atomic number that establishes the uniqueness of an element. Different elements have different atomic numbers.

Arrangement of electrons (electronic structure)

As the distance from the nucleus increases, there are certain regions where electrons are most likely to be found. These regions are called shells or energy levels. The biggest atoms require seven shells to accommodate all their electrons. The shells are numbered 1 to 7 or lettered K to Q outwards from the nucleus. The smaller atoms, of course, require fewer shells. The following table gives the maximum number of electrons which each shell can hold.

SHELL	MAXIMUM NUMBER OF ELECTRONS = $2n^2$ (WHERE n IS SHELL NUMBER)
1 or K	2
2 or L	8
3 or M	18
4 or N	32
5 or O	50
6 or P	72
7 or Q	98

Scientists have not yet made atoms big enough to confirm the maximum number of electrons in the outer three shells. The numbers 50, 72 and 98 are predicted by theory. The electronic structure of gold Au may be represented as

Au 2, 8, 18, 32, 17, 2

which means that the K shell is completely filled with 2 electrons, the L with 8, the M with 18, the N with 32, while the O and P shells are incompletely filled with 17 and 2 electrons respectively. Alternatively, the structure can be shown diagrammatically as in Fig. 4.2.

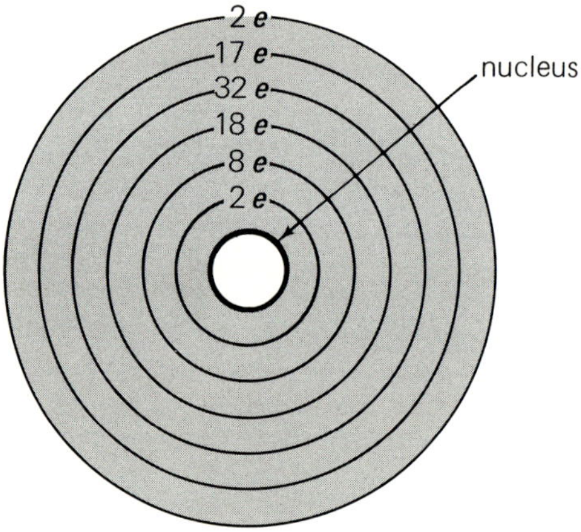

Figure 4.2
Gold atom, showing central nucleus surrounded by electron shells.

EFFECT OF ELECTRONIC STRUCTURE ON PROPERTIES

Similar elements have similar electronic structures meaning that they contain the same number of electrons in their outermost shells which are called the valence shells. Thus chlorine and bromine are similar because they have the electronic structures.

Cl 2, 8, 7 Br 2, 8, 18, 7

Very similar elements will have the same number of electrons in their last two shells. Sodium and potassium have the arrangements

Na 2, 8, 1 K 2, 8, 8, 1

and so their properties are very alike.

PERIODIC TABLE OF THE ELEMENTS

If the elements are arranged in order of increasing atomic number, then at definite intervals, or periods, similar elements occur. In the first part of the table this occurs with every eighth element, in the middle part with every eighteenth and in the latter part with every thirty-second. The reason for this is that after these intervals the elements have similar electronic structures, for example, calcium, Ca with shell structure 2, 8, 8, 2 occurs eight elements after magnesium, Mg with shell structure 2, 8, 2; arsenic, As with shell structure 2, 8, 18, 5 comes eighteen elements after phosphorus, P with shell structure 2, 8, 5; lead, Pb with shell structure 2, 8, 18, 32, 18, 4 is thirty-two elements after tin, Sn with shell structure 2, 8, 18, 18, 4. Similar elements are written underneath one another in the same group as shown in Table I.

Hydrogen appears on its own in the table. This is because its atomic structure is unique in that its K shell is incomplete and its nucleus is simply a proton; it contains no neutrons. All the other atoms contain neutrons. Nonmetals are found in the top right of the table; metalloids (elements possessing properties of both metals and nonmetals) lie on a diagonal line from boron to astatine; the rest are metals.

The periodic table is one of the most important classifications in science.

ISOTOPES

All the atoms of a given element contain the same number of protons. However, their nuclei need not contain the same number of neutrons. For example, carbon may exist as three different atom types: one containing six neutrons, another seven and the third eight. All three types contain six protons and therefore have identical chemistry. They can be represented diagrammatically as in Fig. 4.3.

Such atoms as these, which have the same number of protons but a different number of neutrons, occupy the same place in the periodic table and are therefore called **isotopes** (Gr. *iso*, same, +Gr. *topos*, place). Some isotopes, particularly those produced artificially, are unstable and radioactive.

Table I. The Periodic Table

PERIOD		I ALKALI METALS	II ALKALINE EARTH METALS	TRANSITION METALS	INNER TRANSITION METALS													
1	1 H																	
2		3 Li	4 Be															
3		11 Na	12 Mg															
4		19 K	20 Ca	21 Sc														
5		37 Rb	38 Sr	39 Y														
6		55 Cs	56 Ba	57 La	58 Ce	59 Pr	60 Nd	61 Pm	62 Sm	63 Eu	64 Gd	65 Tb	66 Dy	67 Ho	68 Er	69 Tm	70 Yb	71 Lu
7		87 Fr	88 Ra	89 Ac	90 Th	91 Pa	92 U	93 Np	94 Pu	95 Am	96 Cm	97 Bk	98 Cf	99 Es	100 Fm	101 Md	102 No	103 Lw

metal   metalloid   nonmetal

GROUPS

←— TRANSITION METALS —→

									III	IV	V	VI CHALCOGENS	VII HALOGENS	0 RARE GASES
														2 He
									5 B	6 C	7 N	8 O	9 F	10 Ne
									13 Al	14 Si	15 P	16 S	17 Cl	18 Ar
22 Ti	23 V	24 Cr	25 Mn	26 Fe	27 Co	28 Ni	29 Cu	30 Zn	31 Ga	32 Ge	33 As	34 Se	35 Br	36 Kr
40 Zr	41 Nb	42 Mo	43 Tc	44 Ru	45 Rh	46 Pd	47 Ag	48 Cd	49 In	50 Sn	51 Sb	52 Te	53 I	54 Xe
72 Hf	73 Ta	74 W	75 Re	76 Os	77 Ir	78 Pt	79 Au	80 Hg	81 Tl	82 Pb	83 Bi	84 Po	85 At	86 Rn

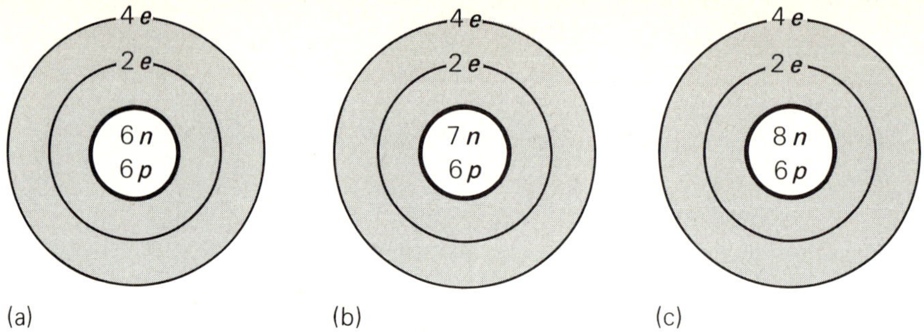

Figure 4.3
Carbon isotopes: (a) carbon 12; (b) carbon 13; (c) carbon 14.

MASS NUMBER

The number of nucleons, the total number of protons and neutrons, in an atom is called its mass number. The mass number of the carbon isotope shown in Fig. 4.3(a) is therefore 12 and it is consequently known as carbon-12, usually written ^{12}C. The other isotopes shown in Fig. 4.3 (b) and (c) are ^{13}C and ^{14}C, respectively.

5

Bonds Between Atoms

In all substances, except the rare gases, the atoms are held together by strong forces which we call *chemical bonds* or simply *bonds*. The properties of a substance are largely determined by the bonds which hold its atoms together and the study of the effect of bonding on properties is really what chemistry is all about.

BONDS BETWEEN ATOMS

Why do atoms form bonds? They do so to become more stable. Experiment has shown that atoms achieve stability by acquiring a certain number of electrons in their outermost shells. The rare gases, for example, already have a stable number of electrons in their outer shells. So atoms of the rare gases (helium, neon, argon, krypton, xenon, and radon) show very little tendency to form bonds; they are very unreactive. The stable number of electrons in the outer shell for helium is 2 and for the rest of the rare gases is 8. When the atoms of other elements form bonds they acquire stable numbers of electrons in their outer shell. Examples of such numbers are 2, 8, 10 and 14.

FORMATION OF BONDS

There are four main types of bonds:

1. Ionic bond
2. Covalent bond
3. Metallic bond
4. Hydrogen bond

and we shall consider, in turn, how each is formed.

Ionic bond

This is formed by an atom losing electron(s) to another atom. Let us look at an example.

Na	2,8,1	+ Cl	2,8,7	$\longrightarrow$	$Na^{\oplus}$	2,8	+ $Cl^{\ominus}$	2,8,8
11*p*	11*e*	17*p*	17*e*		11*p*	10*e*	17*p*	18*e*
SODIUM ATOM		CHLORINE ATOM			SODIUM ION		CHLORINE ION	

The sodium atom loses an electron to the chlorine atom, resulting in the formation of a positive sodium ion $Na^{\oplus}$ and a negative chlorine ion $Cl^{\ominus}$. A glance at the numbers of electrons and protons in each ion should satisfy you that the sodium ion has a net charge of +1 and the chlorine ion a net charge of −1. The bond between these two ions is simply the mutual attraction of oppositely charged particles. This type of bond is called an *ionic bond* and sodium chloride is said to be an ionic compound. You will notice that each ion in the above example has a stable octet of electrons in its outer shell.

Another reaction involving electron transfer is that between calcium and hydrogen resulting in the formation of the ionic compound, calcium hydride.

Ca	2,8,8,2	+	2 H	1	$\longrightarrow$	$Ca^{2\oplus}$	2,8,8	+	2 $H^{\ominus}$ 2
20*p*	20*e*		1p	1e		20*p*	18*e*		1p 2e
CALCIUM ATOM			TWO HYDROGEN ATOMS			CALCIUM ION			TWO HYDROGEN IONS

A calcium atom must lose two electrons in order to acquire a stable octet in its outer shell and a hydrogen atom must gain one electron to obtain the stable doublet of helium. Two hydrogen atoms are therefore needed to take up the two "surplus" electrons of calcium. The upshot is the formation of one positive calcium ion, $Ca^{2\oplus}$ and two negative hydrogen ions, $H^{\ominus}$. The valence of an element is the number of charges carried by each of its ions. Calcium, therefore, has a valence of 2 and is said to be *divalent,* whereas sodium, chlorine, and hydrogen all have a valence of 1 and are said to be *univalent.* You will observe that the valences of calcium and sodium are determined by the number of electrons *in* their outer shell, whereas the valences of hydrogen and chlorine are determined by the number of electrons *needed to complete* their outer shell.

Covalent bond

This is formed by two atoms sharing a pair of electrons. An example of this is

HYDROGEN ATOM CHLORINE ATOM HYDROGEN CHLORIDE MOLECULE

The symbol "·" represents an electron in the hydrogen atom and the symbol "x" an electron in the chlorine atom. Only the electrons in the outermost shells are shown. The shared pair of electrons (shown ringed like this ⓧ) is reckoned as being in the outermost shell of each atom so that the hydrogen atom now has a stable doublet and the chlorine a stable octet of electrons.

The bond between these two atoms is the attraction which the electron pair has for the positive nuclei of hydrogen and chlorine. A bond of this type is called a *single covalent bond* or an *electron-pair bond*. It is often represented by a line joining the two symbols

H—Cl

A compound in which all the bonds are of this type is called a covalent compound. The valence of an element in such a compound is given by the number of bonds it forms, so that hydrogen and chlorine are again seen to be univalent as they are in ionic compounds.

Other examples are

abbreviated to O═O

OXYGEN ATOMS OXYGEN MOLECULE

Two pairs of electrons are shared by the oxygen atoms and the bond between them is called a *double bond*.

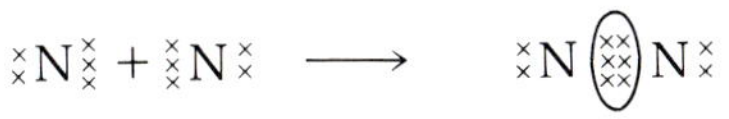

abbreviated to N≡N

NITROGEN ATOMS NITROGEN MOLECULE

The bond between the nitrogen atoms is called a *triple bond* because they share three pairs of electrons.

In the previous three examples each atom contributes one electron to the shared pair. In some compounds both electrons in the shared pair are provided by one of the atoms only. Such an electron pair bond is called a *coordinate* covalent bond. It is represented by an arrow pointing from the donor atom to the acceptor atom.

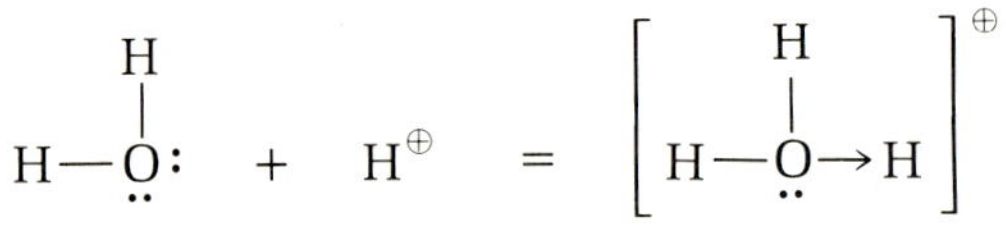

WATER HYDROGEN ION HYDROXONIUM ION

Metallic bond

As its name suggests, this bond unites atoms in a metal. Atoms of all metals need to lose electrons to acquire a stable number of electrons in their outermost shells. The shedding of electrons results in the formation of positive ions. These ions are arranged in a regular fashion in the metal and the spaces between them are occupied by the liberated electrons, as illustrated in Fig. 5.1. The electrons attract the oppositely charged ions which are therefore kept in position. The electrons act rather like a cement.

Hydrogen bond

When hydrogen combines with oxygen, nitrogen, or fluorine, these elements draw the electrical charge, or the electron-pair, away from the hydrogen, producing a slight positive charge on the hydrogen atom. For example, in hydrogen fluoride one end of the molecule is positively charged and the other negatively charged.

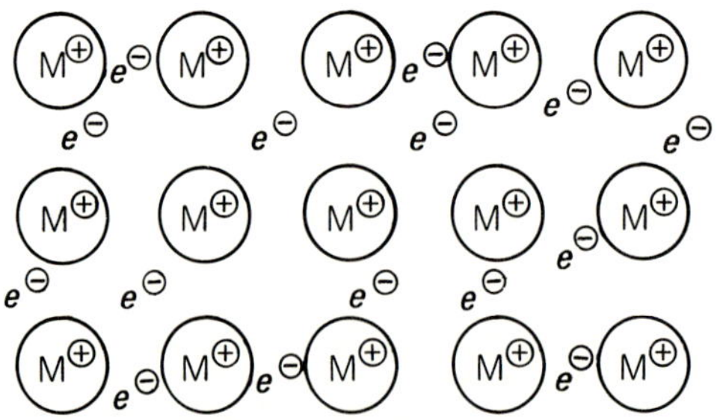

Figure 5.1
Metallic bonding. The metal ions are represented by $M^{\oplus}$ and electrons by $e^{\ominus}$.

$\overset{\delta+}{H} : \overset{\delta-}{F}$ — δ^+ represents a slight positive charge and δ^- a slight negative charge

So the hydrogen is attracted to the negatively charged fluorine atom (in fact, to one of its nonbonding electron-pairs) in another hydrogen fluoride molecule.

$$\overset{\delta+}{H} \quad :\overset{\delta-}{F} \quad \overset{\delta+}{H} \quad :\overset{\delta-}{F}$$

This is an electrostatic bond called the hydrogen bond. It is this type of bond which causes molecules of water to join together, so that water behaves as if it had much bigger molecules than just H_2O, which would be a gas at room temperature.

OXIDATION AND REDUCTION

A great many reactions involve either electron transfer or unequal sharing of electrons. These reactions are grouped together by chemists under the general heading of *redox reactions*. A reaction which involves the loss of electrons is called **oxidation.** The loss may be complete, as by sodium in its reaction with fluorine.

$$Na\cdot + \cdot\ddot{\underset{\cdot\cdot}{F}}: \longrightarrow Na^{\oplus} + :\ddot{\underset{\cdot\cdot}{F}}:^{\ominus}$$

or partial, as by hydrogen in its reaction with fluorine which we have just encountered.

$$H\cdot + \cdot\ddot{\underset{\cdot\cdot}{F}}: \longrightarrow \overset{\delta+}{H} : \overset{\delta-}{\ddot{\underset{\cdot\cdot}{F}}}:$$

The sodium and hydrogen in these reactions are said to be *oxidized* by the fluorine which is called an *oxidizing* agent.

The reverse of oxidation, the complete or partial gain of electrons, is known as **reduction.** The fluorine is *reduced* by sodium in the first reaction and by hydrogen in the second. Sodium and hydrogen are the *reducing* agents in these reactions.

You have no doubt realized that you cannot have oxidation without reduction, because if a particle is to lose an electron there must be another ready to receive it. Reactions involving these two important processes are therefore called *red*uction–*ox*idation reactions, usually abbreviated to *redox* reactions.

CONFIGURATIONS OF ATOMS BONDED TOGETHER

There are two main ways in which atoms can arrange themselves when they bond together. They can form either *molecules* or *giant structures*.

Figure 5.2
Diamond—covalent giant structure. Each carbon atom is joined by covalent bonds to four adjacent carbon atoms.

Molecules, as we have seen, are groups of definite numbers of atoms. The atoms in molecules are always joined together by covalent bonds. Gases, liquids, and low melting-point solids have molecular structures.

In high melting-point solids, however, the atoms or ions bond together not as discrete molecules but as a giant structure of no definite size. When it forms it keeps extending until it is stopped by the sides of the vessel or by a similar giant structure growing towards it. Bonding in

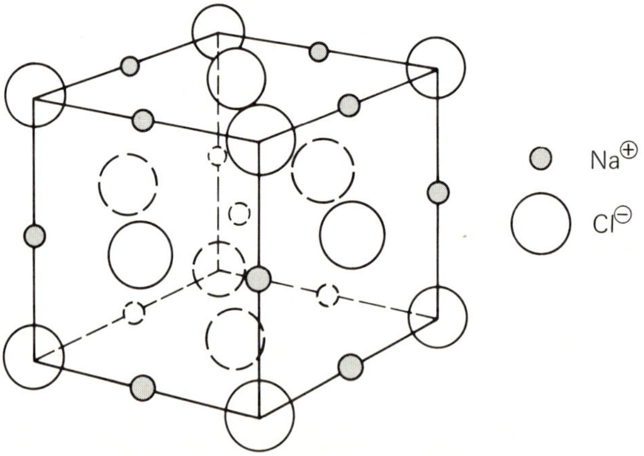

Figure 5.3
Sodium chloride—ionic giant structure. The cube is drawn to show the pattern. The structure is held together by the attraction between positive $Na^{\oplus}$ and negative $Cl^{\ominus}$ ions.

giant structures can be either covalent, as in diamond (Fig. 5.2) or ionic, as in sodium chloride (Fig. 5.3).

Compounds with giant structures are represented by an *empirical formula* (a molecular formula is not appropriate as there are no molecules present). An empirical formula is the simplest possible formula and merely states the simplest atomic ratio of the elements in the compound. Thus, NaCl is the empirical formula of sodium chloride.

BONDS DURING CHEMICAL CHANGES

What happens to the bonds between atoms in a chemical change? The bonds in the starting material must be broken so that the atoms can rearrange themselves to make new substances. New bonds must be made between the atoms before they can produce the new substances.

We always have to supply energy in the form of heat, light or electricity to break bonds; energy is always given out when bonds are made. If the energy given out is less than the energy which we supply, the reaction is *endothermic* and if the energy given out is in excess, it is *exothermic*. An exothermic reaction giving out a great deal of energy can be explosive such as hydrogen and oxygen combining under certain conditions to form water.

6

Counting Atoms and Molecules

It is not possible to weigh a single atom or even a single molecule on a balance, but it has been possible to determine the number of particles in a given weight of substance. Thus, 12 g of carbon-12 contains 6.02×10^{23} atoms or 602,000,000,000,000,000,000,000 atoms. This enormous number is called one **mole.**

The weights containing a mole of particles of some other substances are given in Table II.

We can therefore count particles by weighing them. For example, if we weigh 44 g of carbon dioxide, then we know that we have a mole of carbon dioxide molecules or 6.02×10^{23} molecules. If we want $1\frac{1}{2}$ moles of these molecules, we must weigh out $1\frac{1}{2} \times 44$ g or 66 g of carbon dioxide.

ATOMIC WEIGHT OF AN ELEMENT

This is the number of grams of the element that contains one mole of atoms. Thus, carbon-12 has an atomic weight of 12. This statement is

Table II. Weights of Moles

SUBSTANCE	TYPE OF PARTICLE	NUMBER OF GRAMS CONTAINING A MOLE OF PARTICLES
hydrogen	atom	1
hydrogen	molecule	2
oxygen	atom	16
oxygen	molecule	32
carbon dioxide	molecule	44
sodium chloride	pair of $Na^{\oplus}$ and $Cl^{\ominus}$ ions	58.5

usually abbreviated to $^{12}C = 12$. If you look at Table II you will see that the atomic weights of hydrogen and oxygen are 1 and 16, respectively; in short H = 1 and O = 16.

FORMULA WEIGHT OF A SUBSTANCE

This is the number of grams of the substance that contains one mole of particles. These particles are molecules if the substance has a molecular structure and then the term "molecular weight" is usually used instead of "formula weight." Table II tells us that the formula weight of sodium chloride is 58.5.

Question: *What are the molecular weights of hydrogen, H_2, oxygen, O_2, and carbon dioxide, CO_2?*

Answer: hydrogen, 2; oxygen, 32; carbon dioxide, 44.

Calculation of formula weights

How do we calculate formula weights? The **formula weight** of a substance is easily calculated from its formula and the atomic weights of the elements in the substance. Let us consider as an example the formula weight of carbon dioxide.

The formula of carbon dioxide is CO_2. This tells us that

1 molecule of carbon dioxide contains: 1 carbon atom and 2 oxygen atoms

∴1 mole of carbon dioxide molecules contains: 1 mole of carbon atoms and 2 moles of oxygen atoms

$\therefore$ 1 mole of
carbon dioxide molecules contains: 12 g of carbon and 2×16 g of oxygen ($C = 12$; $O = 16$)

$\therefore$ 1 mole of
carbon dioxide molecules weighs: $(12 + 2 \times 16)\ \text{g} = 44$ g

Therefore the formula weight (or molecular weight) of carbon dioxide = 44. In general, then, the formula weight is obtained by adding up the atomic weights as indicated by the formula. Thus

$$\begin{Bmatrix}\text{formula}\\ \text{weight of calcium}\\ \text{carbonate}\end{Bmatrix} = \begin{Bmatrix}\text{1 atomic}\\ \text{weight of}\\ \text{calcium}\end{Bmatrix} + \begin{Bmatrix}\text{1 atomic}\\ \text{weight of}\\ \text{carbon}\end{Bmatrix} + \begin{Bmatrix}\text{3 atomic}\\ \text{weights of}\\ \text{oxygen}\end{Bmatrix}$$

$$CaCO_3 = 40 + 12 + 3 \times 16$$
$$= 100$$

Question *What is the formula weight of sulfuric acid,* H_2SO_4 ($H = 1$; $S = 32$; $O = 16$)?

Answer 98

A gram-formula or g-formula of a substance. This is its formula weight given in grams. Thus, 1 g-formula of calcium carbonate is therefore 100 g, and $\frac{1}{2}$ g-formula of it would be 50 g.

Question *How many grams of carbon dioxide are equivalent to 2 g-formulas?*

Answer 88 g

NOTE. In the case of molecular substances, the term *gram-molecule* (g-molecule) is usually used instead of g-formula.

A gram-atom or g-atom of an element. This is its atomic weight in grams. Thus, 1 g-atom of oxygen is 16 g.

Question *What are $2\frac{1}{2}$ g-atoms of iron,* Fe *expressed in grams* ($Fe = 56$)?

Answer 140 g

NOTE 1. In current usage, a formula may represent either one particle or one mole of particles; that is a gram-formula. Likewise, the symbol of an element can represent either one atom or one gram-atom of it. You must decide from the context which purpose the formula and the symbol are serving.

NOTE 2. It is important to remember that a g-atom and a g-formula contain the same number of particles, that is, one mole. The importance of these quantities will be dealt with in the chapter on equations that follows.

CONCENTRATION OF SOLUTIONS

You may have noticed on the labels of bottles, cryptic statements such as 2 mmol/liter (2mmol/l). What does all this mean? It is simply the chemist's way, a very good way in fact, of denoting the concentration of the solution in the bottle. The term "mmol" is short for millimole which means one-thousandth of a mole (note that milli- represents a thousandth not a millionth). The quantity 2 mmol/liter means two-thousandths of a mole per liter; each liter of the solution contains two-thousandths of a mole of particles of solute. Solute is the name given to the dissolved substance. Or if you prefer, we can say each liter of solution contains two-thousandths of a g-formula of solute.

Another way of denoting concentration is to say that a solution is "so-and-so" **molar.** For example, a solution might be 0.5 molar, which is usually abbreviated to 0.5*M*. This means that it contains 0.5 moles of solute per liter.

A less informative method is to state how many grams of solute are dissolved in each liter. The label on a bottle of saline solution (sodium chloride NaCl solution) might say that it contains 8.5 g/liter. This is less informative because it does not state how many sodium chloride particles are present in one liter; and it is particles rather than weight that the chemist is particularly interested in. However, it is not a difficult job to convert g/liter into moles/liter.

Question *How many moles/liter are there in an 8.5 g/liter saline solution?*

Answer: formula weight of NaCl = 58.5 (see Table II)

$$\therefore 58.5 \text{ g NaCl} = 1 \text{ mole NaCl}$$

$$\therefore 8.5 \text{ g NaCl} = 8.5/58.5 \text{ moles NaCl} = 0.15 \text{ moles NaCl}$$

$$\therefore 8.5 \text{ g NaCl/liter} = 0.15 \text{ moles NaCl/liter} = 0.15M \text{ NaCl}$$

There is still another way of stating a solution's concentration that we should know about. It is really a variation on the last method. Instead of saying how many grams are dissolved in a liter (which is equivalent to 1000 cubic centimeters or 1000 cm^3), we sometimes say how many grams are dissolved in 100 cm^3 and denote the result as a percentage. Thus, the 8.5 g/liter saline solution referred to earlier contains 0.85 g/100 cm^3 and is therefore called a 0.85% saline solution. This has been quite a common way of denoting concentration. However, in medical practice it is now being replaced by SI Units (Système International d'Unités).

SI UNITS (SYSTÈME INTERNATIONAL D'UNITÉS)

In the International System, there is only one base unit for each physical quantity based on the metric system. In this book we are concerned mainly with amounts of substances and with volumes. The pattern of this system is outlined in Table III.

Wherever possible for our purposes, the concentration of substances will be expressed in millimoles per liter, with the units recently used in medical practice given in brackets. In a number of instances, however, for example, where the molecular weight is not known, the concentration is still expressed in terms of the gram, g, or one of its derived units:

milligram = mg = $g \times 10^{-3}$; that is,

$$\frac{g}{1000} \qquad \text{(one-thousandth of a gram)}$$

microgram = μg = $g \times 10^{-6}$; that is,

$$\frac{g}{1{,}000{,}000} \qquad \text{(one-millionth of a gram)}$$

nanogram = ng = $g \times 10^{-9}$; that is,

$$\frac{g}{1{,}000{,}000{,}000} \qquad \text{(one-thousand millionth of a gram)}$$

picogram = pg = $g \times 10^{-12}$; that is,

$$\frac{g}{1{,}000{,}000{,}000{,}000} \qquad \text{(one-million-millionth of a gram)}$$

Table III. SI Units

	BASE UNIT	SYMBOL	DERIVED UNITS	SYMBOL
AMOUNT OF SUBSTANCE	mole	mol	millimole (= mol/1000)	mmol
LENGTH	meter	m	centimeter (= m/100 = $m/10^2$)	cm
			millimeter (= m/1000 = $m/10^3$)	mm
			nanometer (meter/10^9)	nm
VOLUME	cubic meter (This is too big to be practical.)	m^3	cubic centimeter (= milliliter)	cm^3 (ml)
			litre (= 1000 cm^3)	l

7

Chemical Equations

A chemical equation is a shorthand way of representing a reaction using the formulas of the substances concerned. Molecular formulas are used for molecular substances, and empirical formulas are used for those with giant structures.

We have already come across chemical equations in Chapter 5, but we used electronic formulas there in order to show how bonds are formed. Normally such formulas are not used in equations.

The equation for the reaction that occurs when iron is burnt in chlorine gas to form iron(III) chloride (the roman numeral refers to the valence of iron) is

$$2Fe + 3Cl_2 = 2FeCl_3$$

Notice that this equation, like all others, is in accord with Dalton's atomic theory which states that atoms are neither created nor destroyed in normal chemical changes. This is confirmed by experiment. This means that the equation must be *balanced*. By this we mean that on each side of it, there must be the same number of atoms of each element. In the above equation, we observe that there are two iron atoms and six chlorine atoms on each side.

It is wrong to write the equation

$$Fe + Cl_2 = FeCl_3 \qquad \text{(incorrect)}$$

because there are three chlorine atoms on the right-hand side, but only two on the left. This implies that chlorine atoms are created in the reaction and this is not the case.

We might be tempted to balance the equation by writing

$$Fe + Cl_3 = FeCl_3$$

You must resist this temptation! The formula Cl_3 is not the molecular formula of chlorine, and it is wrong to suggest that it is.

How, therefore, do we balance an equation? One of the surest methods is to balance the number of atoms of each element in turn by moving from left to right across the left-hand side of the equation. In our example, the first element on the left-hand side is iron and there is one atom of it. There is also one iron atom on the right-hand side. Therefore, the iron atoms are balanced. The next element as we move to the right is chlorine and on the left-hand side there are two chlorine atoms that clearly do not balance the three chlorine atoms on the right-hand side. If there are three chlorine atoms at the end of the reaction, there must be three at the beginning. Thus, there must be $1\frac{1}{2}$ molecules of chlorine to start with. The equation is therefore balanced if we write

$$Fe + 1\tfrac{1}{2}Cl_2 = FeCl_3$$

This equation is not acceptable, however, as it is not possible to obtain half a molecule of chlorine under normal conditions. In order to get rid of the fraction from the equation, we must multiply both sides by two. This gives

$$2Fe + 3Cl_2 = 2FeCl_3$$

which is the correct equation for the reaction. However, it is always a wise precaution to check that the equation really is balanced by repeating the balancing process. This is because it sometimes happens that in balancing the atoms of one element, we upset the balance already achieved for the atoms of another element. The following example illustrates this difficulty. Let us consider ammonia reacting with heated copper(II) oxide to form copper, water and nitrogen. We obtain the equation for this reaction by first writing

$$\underset{\text{AMMONIA}}{NH_3} + \underset{\text{COPPER(II) OXIDE}}{CuO} \longrightarrow \underset{\text{COPPER}}{Cu} + \underset{\text{WATER}}{H_2O} + \underset{\text{NITROGEN}}{N_2}$$

The "equals" sign is not used because this equation is not balanced. There is one nitrogen atom on the left, but there are two on the right. We must therefore have two molecules of ammonia on the left-hand side to balance the nitrogen atoms. We obtain the equation for this reaction by first writing

$$2NH_3 + CuO \longrightarrow Cu + H_2O + N_2$$

The next element as we move from left to right is hydrogen. There are six hydrogen atoms on the left-hand side but only two on the right-hand side. In order to have six hydrogen atoms on the right, we must have three molecules of water. Therefore we write

$$2NH_3 + CuO \longrightarrow Cu + 3H_2O + N_2$$

The next element is copper and its atoms are balanced, one on each side. Finally, there are the oxygen atoms to consider with one on the left and three on the right. If we end up with three, we must start with three; there must be three particles of copper(II) oxide to begin with. Thus

$$2NH_3 + 3CuO \longrightarrow Cu + 3H_2O + N_2$$

If we check that the equation is balanced by repeating the above process from left to right, we find that the copper atoms are no longer balanced. This is easily corrected by having three atoms of copper on the right-hand side. Thus

$$2NH_3 + 3CuO = 3Cu + 3H_2O + N_2$$

An "equals" sign can now be used since a final check shows that the equation is now balanced.

Sometimes we indicate in an equation whether the substances are solid (s), liquid (l), or gaseous (g), or aqueous, dissolved in water, (aq). The equation

$$\underset{(s)}{Na_2CO_3} + \underset{(aq)}{H_2SO_4} = \underset{(aq)}{Na_2SO_4} + \underset{(l)}{H_2O} + \underset{(g)}{CO_2}$$

tells us that the solid sodium carbonate, Na_2CO_3 reacts with aqueous sulfuric acid, H_2SO_4 to form aqueous sodium sulfate, Na_2SO_4, liquid water, H_2O, and carbon dioxide gas, CO_2.

We mentioned in the last chapter that in current practice a formula is often taken to represent one mole of particles. It should be appreciated, therefore, that an equation tells us not merely what substances are involved in a reaction, but also their relative amounts. For example, the equation

$$\underset{(s)}{Fe_2O_3} + \underset{(g)}{3CO} = \underset{(s)}{2Fe} + \underset{(g)}{3CO_2}$$

tells us that 1 mole of iron(III) oxide, Fe_2O_3 reacts with 3 moles of carbon monoxide, CO to give 2 moles of iron atoms Fe and 3 moles of carbon dioxide, CO_2. These quantities are, of course, relative. The actual amounts, for example, might be $\frac{1}{2}$ mole of iron(III) oxide, $1\frac{1}{2}$ moles of carbon monoxide, 1 mole of iron and $1\frac{1}{2}$ moles of carbon dioxide. The mole quantities can easily be converted into grams if we know the formula weights of the substances.

METHOD FOR DETERMINING EQUATIONS

We referred to molar solutions in Chapter 6. These can be used to determine the equation for a reaction that occurs in solution. For example, experiment shows that 1 liter of *M* H_2SO_4, or molar sulfuric acid solution, reacts completely with 2 liters of *M* NaOH. In other words, 1 mole of sulfuric acid, represented by its formula H_2SO_4, reacts with 2 moles of sodium hydroxide represented by 2NaOH. Thus, the left-hand side of the equation must be

$$H_2SO_4 + 2NaOH$$

We can complete the equation provided that we know what substances are formed. In this case, they are sodium sulfate and water. The equation must be

$$H_2SO_4 + 2NaOH = \underset{\text{SODIUM SULFATE}}{Na_2SO_4} + H_2O$$

Not only have we deduced the equation, but also the formula for sodium sulfate. It must be Na_2SO_4 in order to balance the equation.

Experimentation is done by using the apparatus sketched in Fig. 7.1. It is possible to determine the volume of *M* NaOH that is required to react completely with 10 cm^3 of *M* H_2SO_4 (we use 10 cm^3 as it is extravagant and unwieldy to use a liter). Our determination is easily made since there are certain substances, such as screened methyl orange and phenolphthalein, that obligingly change color when all the acid has reacted.

Figure 7.1
Titration, using burette and flask.

These substances are, not surprisingly, called *indicators*. It is found that 10 cm^3 of M H_2SO_4 requires 20 cm^3 of M NaOH for a complete reaction. Therefore, 1 liter of M H_2SO_4 will react completely with 2 liters of M NaOH, which enables us to determine the equation as explained above.

The use of a burette in this way (Fig. 7.1) is known as *titration*, and by convention we say that one solution is titrated against the other.

REVERSIBLE REACTIONS

Sometimes we use "double arrows" in an equation.

$$\underset{\text{NITROGEN}}{N_2} + \underset{\text{HYDROGEN}}{3H_2} \rightleftharpoons \underset{\text{AMMONIA}}{2NH_3}$$

These double arrows tell us that the reaction is reversible; by this we mean that it "goes both ways." In the example given, nitrogen and hydrogen combine to form ammonia, called the forward reaction; and ammonia also decomposes to give nitrogen and hydrogen, called the reverse reaction. It would therefore appear impossible to make ammonia from its constituent elements. This is not so, however, for the following reason.

The rate at which substances react together depends on their concentration. The greater the concentration, the faster the substances react since there is a greater chance of the particles colliding. This means that when ammonia is first produced, it decomposes slowly because its concentration is low. Conversely, nitrogen and hydrogen react relatively fast at the outset to form ammonia because their concentration is high. As time goes on the concentration of ammonia increases and so the ammonia decomposes faster.

Meanwhile, the ammonia is being produced more slowly, as the concentrations of nitrogen and hydrogen are decreasing. Eventually, ammonia decomposes as fast as it is being formed. That is, the speed of the forward reaction balances the speed of the reverse reaction. Beyond this balancing point, it is not possible to make any more ammonia under these conditions, but at least some ammonia can be made. When the speeds of the forward and reverse reactions become equal (or balanced), we say that the system is in *dynamic equilibrium*. We refer to this equilibrium as dynamic because the forward and reverse reactions continue after the equilibrium has been reached, but their speeds remain equal.

There are many more chemical equations to be looked at as we continue to explore the field of chemistry. These equations should present us with no difficulty provided that we have read and examined this chapter carefully.

8

Combustion and Respiration

Antoine Lavoisier, the French chemist, whose brilliant career was truncated by the guillotine, realized that combustion provided the clue to the chemistry of respiration. Both processes are characterized by the release of energy. In the combustion of a fire the energy is released mainly in the form of heat and the rate of release is very rapid. In respiration only a small part of the energy is released as heat, the greater part being stored in energy-rich compounds that release energy when it is required for the various forms of cellular activity. This latter process is under firm control. Before delving into the differences between these two processes, let us first investigate their similarities.

COMBUSTION

EXPERIMENT 1

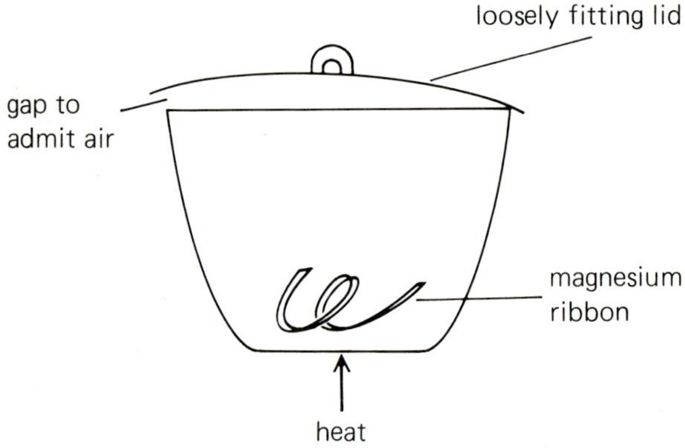

Figure 8.1
Apparatus for Experiment 1 (combustion).

Method

A porcelain crucible containing a piece of magnesium ribbon is weighed, together with the crucible lid. The crucible is then heated for a few minutes as shown in Fig. 8.1, so that the magnesium burns. The crucible and the white ash that remains are allowed to cool and are then weighed together with the lid.

Observation

The white ash weighs more than the magnesium.

Conclusion

The magnesium has gained "something."

Where does the "something" come from?

One possible source is the air. We must now test our hunch in the next experiment.

EXPERIMENT 2

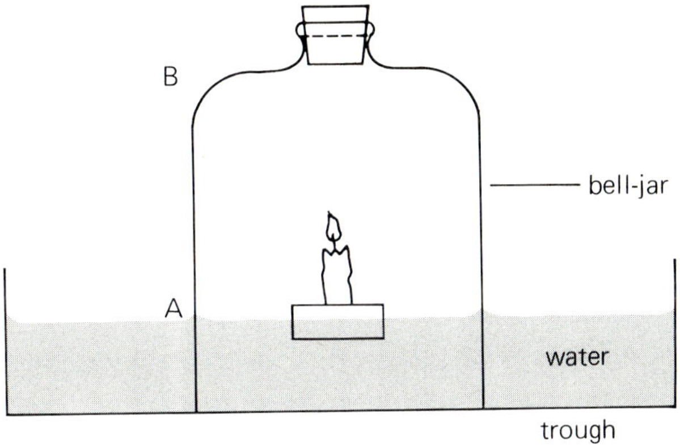

Figure 8.2
(a) Apparatus for Experiment 2 (combustion).

Method

A burning candle is placed on a block of wood which floats on water in a trough. A bell-jar is now placed over the candle and then stoppered (Fig. 8.2(a)).

Observations

i. The candle soon stops burning.
ii. The water rises a little inside the bell-jar (Fig. 8.2(b))

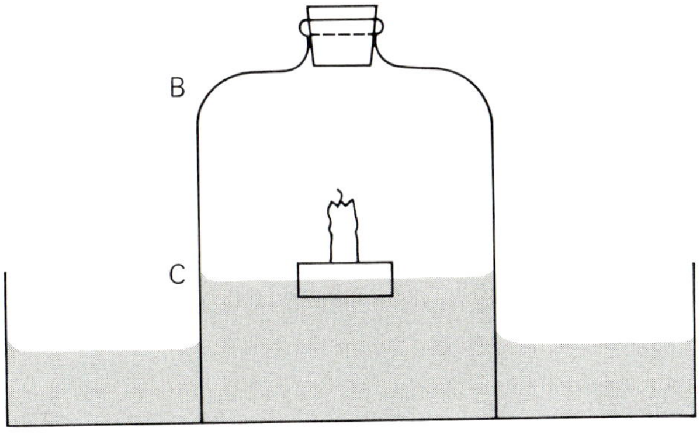

Figure 8.2
(b) Water rising in bell-jar.

Conclusion

Part of the air is used up during burning; our hunch was right! We call this part of the air oxygen. The rest is mainly nitrogen.

Further support for our conclusion

To prove beyond doubt that a burning substance gains oxygen from the air, we must recover the oxygen from the product of burning. This is usually not an easy thing to do. It was first done in the 18th century by Lavoisier. He recovered oxygen from the red powder left after mercury is burnt by heating the powder strongly. This decomposed it into mercury and a colorless, odorless gas. He then proved that this gas was the one taken out of the air by the burning mercury; in other words he proved it to be oxygen.

Lavoisier then put forward the *theory of combustion* which we hold today. His theory states that when a substance burns in air, it combines with oxygen. The energy given out in this chemical reaction appears as heat and if there is a flame or glow, some of the energy is in the form of light as well (a flame is simply hot gas, which emits light). The burning substance combines with oxygen to form an oxide. The white ash formed when magnesium burns in air is magnesium oxide and the equation for the reaction is

$$2Mg + O_2 = 2MgO$$

2Mg	O_2	2MgO
MAGNESIUM (s)	OXYGEN (g)	MAGNESIUM OXIDE (s)

The magnesium is *oxidized* to magnesium oxide since it loses electrons to the oxygen.

Some substances leave no visible product of burning; they burn to form colorless gases. Carbon, for example, forms carbon dioxide. In the last experiment, the hydrocarbons (compounds containing carbon and hydrogen only) which make up the candle, burn to form carbon dioxide and water vapor.

RESPIRATION

EXPERIMENT 1

Breathe on a cold, dry metal spoon.

Observation

The spoon becomes dull and wet. The liquid can be shown to be water.

Conclusion

Water vapor is produced when we breathe. The vapor condenses on the cold spoon.

EXPERIMENT 2

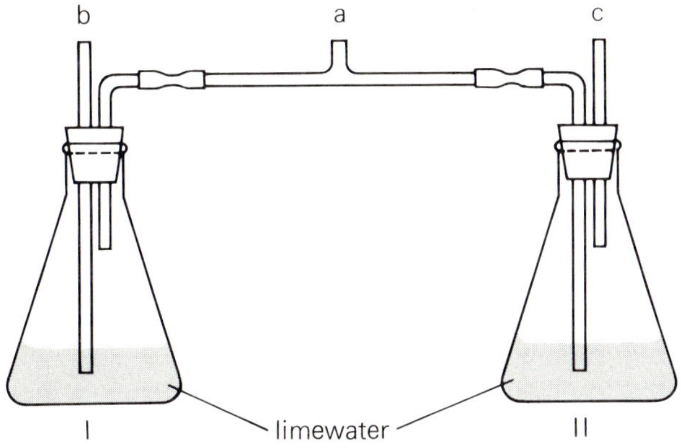

Figure 8.3
Apparatus for Experiment 2 (respiration).

Using the apparatus shown in Fig. 8.3 we can identify another product of breathing.

Method

One breathes through tube a, closing tube c with the finger when breathing in and closing tube b when breathing out. This ensures that the inhaled air passes through the limewater in flask I and the exhaled air through the limewater in flask II.

Observation

The limewater in flask II soon turns milky. This is the test for carbon dioxide.

Conclusion

Flask I is used because air contains a small proportion of carbon dioxide. This is absorbed by the limewater in flask I. Milkiness in flask II can therefore be caused only by carbon dioxide in exhaled air.

HUMAN PRODUCTION OF CARBON DIOXIDE AND WATER VAPOR

Why are carbon dioxide and water vapor produced when we breathe? The oxygen, which we take in when we breathe, reacts with the sugars and starches which the body gets from food. The products of these chemical reactions are carbon dioxide, water vapor and energy.

$$\underset{\text{GLUCOSE (A SUGAR)}}{C_6H_{12}O_6} + 6O_2 = 6CO_2 + 6H_2O + \text{energy}$$

The energy of this reaction keeps us alive. It enables all the muscles of the body to work and keeps the temperature of the body virtually constant at 37°C, the temperature at which the body functions best.

Thus, when we breathe, oxygen from the air reacts with substances in our bodies to produce oxides and energy. The end products are the same as those produced when these substances burn in air. The energy produced is also the same, 2831 kilojoules for every mole of glucose. It is the steps by which the energy is released that are different.

OXYGEN IN THE HUMAN BODY

When we inhale, air passes down our air passages, which branch into progressively smaller and smaller tubes, until it reaches small blind sacs called alveoli (Fig. 8.4(a)).

These alveoli have very thin walls through which dissolved oxygen can readily pass to enter the blood flowing through the numerous capillaries in the alveolar walls (Fig. 8.4(b)).

The movement of air down the larger tubes is mainly due to the muscles of respiration and the natural elasticity of the lungs. The movement of the air gases through the smallest airways and particularly through the walls of the alveoli is largely due to the process called diffusion. This, as you will no doubt remember, is the natural tendency of gases to spread throughout any space they occupy, due to their molecules being in constant motion.

When the air gases enter the lung capillaries, the dissolved oxygen is immediately taken up by the hemoglobin in the red cells. Hemoglobin before it combines with the oxygen is called reduced hemoglobin and is blue in color. After combination with oxygen it becomes bright red oxyhemoglobin. The oxyhemoglobin in the red cells is carried by the circulation to the tissues where it readily dissociates into reduced hemoglobin and oxygen.

The combination of hemoglobin with oxygen is a reversible reaction, represented by the double arrows in the following equation.

$$\text{hemoglobin} + \text{oxygen} \rightleftharpoons \text{oxyhemoglobin}$$

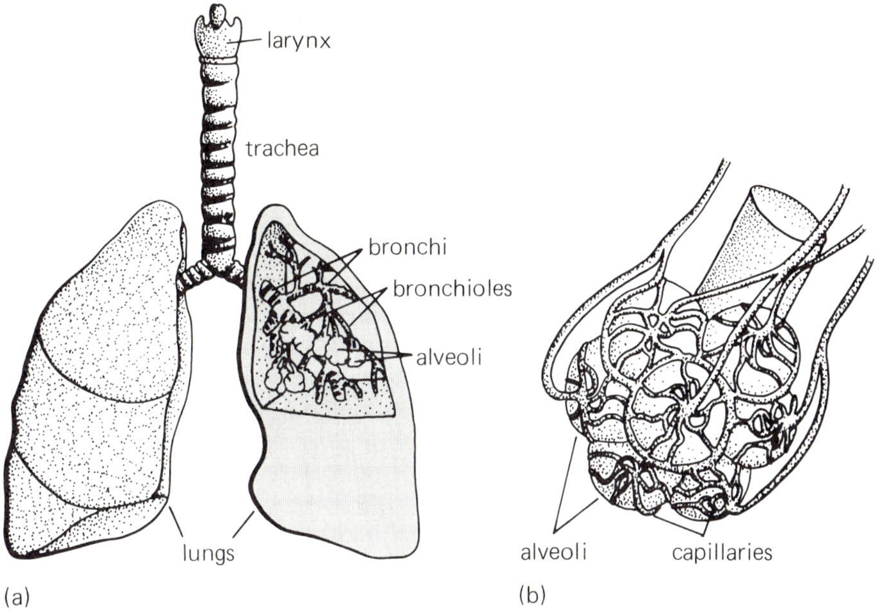

Figure 8.4
(a) Air passages leading to alveoli; (b) capillary network in alveolar walls.

The arrows indicate the dynamic equilibrium between the two sides of the equation. If more oxygen is added to the left-side it promotes the formation of more oxyhemoglobin on the right-side to restore the balance. If oxygen is removed from the left-side, the balance is restored by oxyhemoglobin dissociating to release more oxygen.

In the lungs the oxygen is practically at atmospheric pressure and there is a correspondingly high oxygen tension, by this we mean pressure in fluid, in the lung capillaries. These conditions promote the formation of oxyhemoglobin and the reaction proceeds from left to right as

$$\text{hemoglobin} + \text{oxygen} \longrightarrow \text{oxyhemoglobin}$$

When the blood reaches the tissue capillaries the conditions are different, especially if the tissues are active, for example in a contracting muscle. Active tissues use up oxygen and so the oxygen tension in the tissue capillaries is lower than in the lungs. The low oxygen tension causes the reaction to proceed from right to left.

$$\underset{\text{(BLUE)}}{\text{hemoglobin}} + \text{oxygen} \longleftarrow \underset{\text{(RED)}}{\text{oxyhemoglobin}}$$

The bright red oxyhemoglobin dissociates, giving up its oxygen and becoming reduced hemoglobin which gives venous blood its bluish

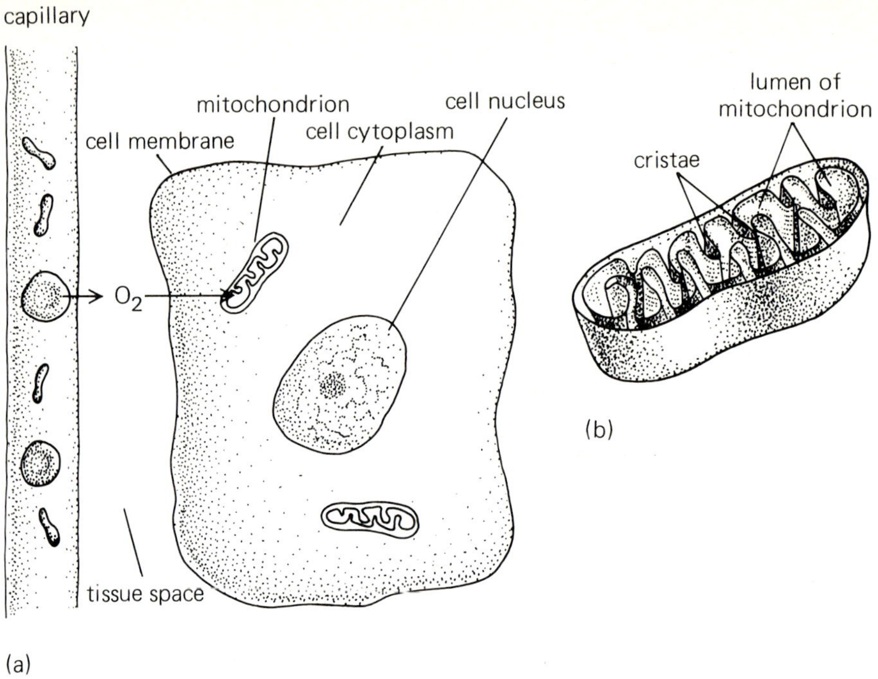

Figure 8.5
(a) Diffusion of oxygen from red cells to mitochondria of tissue cells; (b) mitochondrion cut across to show internal structure (Fig. 28.5).

color. Where tissue is active there is a raised temperature and increased acidity. Both these factors have been shown by experiment to assist in promoting the release of oxygen from oxyhemoglobin. The released oxygen dissolves in the plasma and tissue fluid, through which it travels by diffusion to reach the cells. The cell wall allows it to pass through into the cytoplasm where it finally reaches the mitochondria (Fig. 8.5a)). These are organelles, small organs, within the cell which are characteristically sausage-shaped and incompletely divided by cristae into a series of communicating compartments (Fig. 8.5(b)).

It is in the mitochondria that the actual combination of oxygen with sugar takes place. If the combination occurred all at once the amount of energy released would be so great that the cell would be destroyed. Instead, the reaction takes place in a series of small steps, releasing a small amount of energy at each step. The structure of the mitochondria appears very suitable for a series of chemical reactions occurring in such steps. The energy released is used to build special chemical molecules providing short term stores of energy, which can be used by tissues such as muscles or brain cells as required. These energy-rich molecules are called ATP, *a*denosine *trip*hosphate. Without a continual supply of

ATP the cells would stop functioning and we would die. Respiration is essential to maintain this supply.

REMOVAL OF CARBON DIOXIDE FROM THE HUMAN BODY

When the sugar (glucose) has been oxidized in the cell, carbon dioxide and water are formed.

$$C_6H_{12}O_6 + 6O_2 = 6CO_2 + 6H_2O + \text{energy}$$

Carbon dioxide is a gas, although no gas is actually produced because it dissolves in water to form carbonic acid.

$$CO_2 + H_2O \longrightarrow \underset{\text{CARBONIC ACID}}{H_2CO_3}$$

This would be poisonous to the cells if it were allowed to accumulate. But it rapidly diffuses out of the cells into the surrounding tissue fluid and from these into the capillaries (Fig. 8.6).

Once it reaches the blood the carbonic acid is partly neutralized by the sodium carbonate in the blood stream. Bicarbonate of soda is produced.

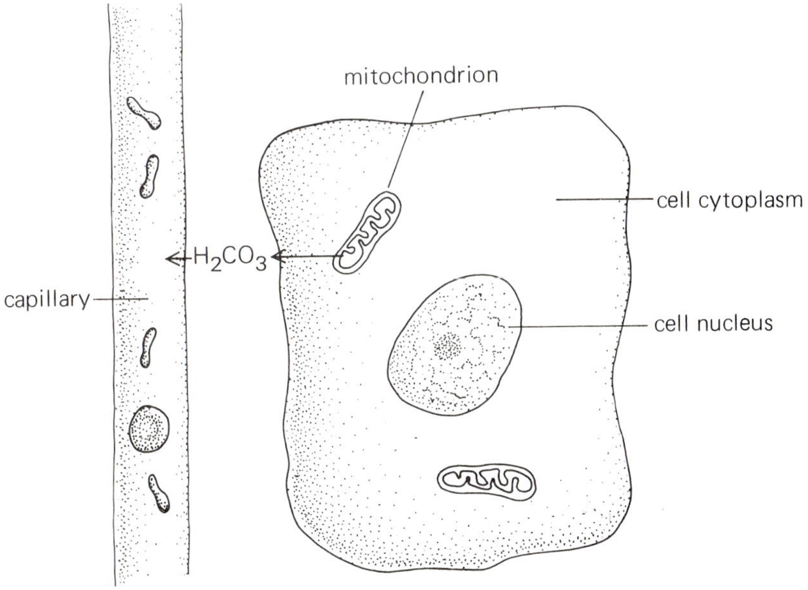

Figure 8.6
Diffusion of carbonic acid, H_2CO_3 from cell to capillary.

$$\underset{\text{SODIUM CARBONATE}}{Na_2CO_3} + \underset{\text{CARBONIC ACID}}{H_2CO_3} \rightleftharpoons \underset{\substack{\text{SODIUM BICARBONATE} \\ \text{(BICARBONATE OF SODA)}}}{2NaHCO_3}$$

This is another reversible reaction, as indicated by the arrows pointing in each direction. In the tissues, when cell respiration is continually producing more carbon dioxide there is a relatively high concentration of carbonic acid.

The more carbonic acid is produced, the more it combines with sodium carbonate to form bicarbonate of soda.

$$Na_2CO_3 + H_2CO_3 \longrightarrow 2NaHCO_3$$

But when it reaches the lungs the process is reversed. The carbonic acid is now very weak because the lungs are continually removing it and so the bicarbonate of soda readily splits up, or dissociates, to release more carbonic acid.

$$2NaHCO_3 \longrightarrow Na_2CO_3 + H_2CO_3$$

The carbonic acid diffuses out from the lung capillaries into the alveoli, further dissociating into carbon dioxide and water which are breathed out in the exhaled air.

$$H_2CO_3 \longrightarrow H_2O + CO_2$$

It is, in fact, the amount of carbon dioxide dissolved in the blood which is the main factor in controlling the depth and rapidity of respiration, acting through the respiratory center in the brain. The anaesthetist uses carbon dioxide in the operating theater when he wants to stimulate the patient's respiration. Too much carbon dioxide, however, is dangerous as it poisons the respiratory center.

OXYGEN AND NITROGEN IN THE AIR

What are the proportions of oxygen and nitrogen in the air? Approximate values for these can be obtained quite easily from Experiment 2 concerning combustion. The volume of gas in the bell-jar is proportional to the height of the jar above the water surface. The volume of air is proportional to AB (Fig. 8.2(a)) and the volume of nitrogen is proportional to BC (Fig. 8.2(b)). Thus,

$$\frac{\text{volume of nitrogen}}{\text{volume of air}} = \frac{BC}{AB}$$

$$\text{percentage by volume of nitrogen in the air} = \frac{BC}{AB} \times 100$$

Since the volume of a gas varies with the temperature and the pressure, a more accurate result is obtained if BC is measured after (1) allowing the apparatus to cool and (2) equalizing the water levels inside and outside the bell-jar. These two precautions ensure that the original air and the remaining nitrogen are at the same temperature and pressure.

The proportion of nitrogen in the air is found to be about 80%. Therefore, the proportion of oxygen is about 20%.

A more complete analysis of air shows it to be a mixture of a number of gases with the following volume composition:

nitrogen	about 80%
oxygen	about 20%
carbon dioxide	about 0.03%
rare gases (helium, neon, argon, krypton, and xenon)	about 1%
water vapor	variable percentage

PHOTOSYNTHESIS

Why do the proportions of oxygen and carbon dioxide in the air remain constant? All living things breathe, by this we mean take in oxygen from the air and give out carbon dioxide into the air. Fuels, which are compounds of carbon, also replace atmospheric oxygen with carbon dioxide when they are burnt. Why is it then that the oxygen of the air has not been completely replaced by carbon dioxide?

The reason is that green plants, in daylight, take up carbon dioxide from the air and give out oxygen. This compensatory process is known as *photosynthesis* (Gr. *photo*, light, +Gr. *synthesis*, building up). The process is so-called because the green plants, with the aid of light energy, build up sugars and starches from carbon dioxide in the air taken in by the leaves and water in the soil taken up by the roots.

$$\underset{(g)}{6CO_2} + \underset{(l)}{6H_2O} + \text{light energy} = \underset{(aq)}{C_6H_{12}O_6} + \underset{(g)}{6O_2}$$

We notice that photosynthesis is the reverse of the reaction which takes place during respiration. It occurs only in green plants because the green coloring matter, called chlorophyll, is needed to absorb light energy. It is this energy, released from food by respiration, which provides the power house for virtually all life.

9

Electrochemistry

ENERGY CHANGES IN CHEMICAL REACTIONS

Energy is what we and machines need to do work. There are various forms of energy, such as heat, light, electrical, and kinetic energy. The steam engine works only if it receives heat from burning fuel. The electric lawn mower will not work without electrical energy. The moving hammer which knocks a nail into the wall for us, possesses energy simply because it is moving. This type of energy is called kinetic energy. If the hammer were stationary it would have no kinetic energy and would be unable to do work for us.

The *law of conservation of energy* is a fundamental law of nature. It states that provided there is no change in mass, energy cannot be created or destroyed. It can be changed in form, that is, changed into some other kind of energy.

Most chemical reactions give out heat: they are called exothermic (Gr. *exo,* out, + Gr. *therme,* heat). The law of conservation of energy tells us that this heat cannot be created from nothing. Some other form of energy must have been converted into heat. Clearly then, the chemicals formed in the reaction must have less energy than the chemicals from which they were made, so that some chemical energy is changed into heat energy. If a reaction is endothermic, absorbs heat, then during the chemical change heat is converted into chemical energy, so that the chemicals formed have more chemical energy than those which produced them.

In some reactions, chemical energy is changed into light which is also a form of energy. We are all familiar with the light from a coal fire. A more exotic example is the "fire" of the firefly. This is one of the rare examples of the chemical emission of "coldlight." The reverse process, changing light into chemical energy, is the principle underlying photography. Light falls on a photographic film and changes a little silver bromide in the film into silver. The amount of silver formed is proportional to the amount of light falling on the film. Thus, we obtain a silver "image" of the subject photographed.

The outstanding example, however, of the conversion of light into chemical energy is photosynthesis which we met with in the last chapter. Sunlight is converted into chemical energy in green plants, the chemical energy being stored in the sugars and starches which are photosynthesized. Animals obtain all their energy requirements from sugars and starches in plants and from animal proteins which are themselves built up from plant sugars and starches. Fuels, such as oil, gas, and coal are products of animal and vegetable decay. The chemical energy of fuels is therefore derived from the sun. Apart from nuclear energy (which we shall discuss in a later chapter) all energy on earth comes from the sun. It is not surprising that primitive peoples were sunworshippers; perhaps they had more insight into life processes than we give them credit for.

ELECTROCHEMISTRY

This is that branch of chemistry that is concerned with the interconversion of chemical and electrical energy. A device that converts chemical energy into electrical is called a *voltaic cell* and one which does the opposite is called an *electrolytic cell*.

You are, of course, familiar with voltaic cells. The car, flashlight, and hearing-aid batteries are common examples. Electrolytic cells are probably less well known to you, but they are of great industrial importance. Sodium, chlorine, aluminium and certain household bleaches are just a few examples of substances which are produced in electrolytic cells. Copper electroplating is also carried out in this type of cell.

VOLTAIC CELLS

In these cells a chemical reaction is carried out in such a way that chemical energy is converted into electrical energy instead of into heat. The reaction that is used in the car battery is

$$\underset{\substack{\text{(s)}\\ \text{LEAD}}}{Pb} + \underset{\substack{\text{(s)}\\ \text{LEAD}\\ \text{DIOXIDE}}}{PbO_2} + \underset{\substack{\text{(aq)}\\ \text{SULFURIC}\\ \text{ACID}}}{2H_2SO_4} = \underset{\substack{\text{(s)}\\ \text{LEAD}\\ \text{SULFATE}}}{2PbSO_4} + \underset{\text{(l)}}{2H_2O}$$

The equation in ionic form is

$$Pb + PbO_2 + \underset{\substack{\text{HYDROGEN}\\ \text{IONS}}}{4H^{\oplus}} = \underset{\substack{\text{LEAD}\\ \text{IONS}}}{2Pb^{2\oplus}} + 2H_2O$$

This is a redox reaction, as indeed are all voltaic cell reactions. This is evident if we split the reaction into two halves

$$Pb = Pb^{2\oplus} + \underset{\text{(2 ELECTRONS)}}{2e^{\ominus}} \qquad \text{oxidation}$$

$$PbO_2 + 4H^{\oplus} + \underset{\text{(2 ELECTRONS)}}{2e^{\ominus}} = Pb^{2\oplus} + 2H_2O \qquad \text{reduction}$$

Electrons are transferred from the lead to the lead dioxide and hydrogen ions.

If the lead, lead dioxide and acid are mixed in a beaker, the chemical energy is changed into heat. How do we prevent this happening in the battery? The solution is rather ingenious. The lead and lead dioxide are separated as shown in the schematic representation of the cell in Fig. 9.1.

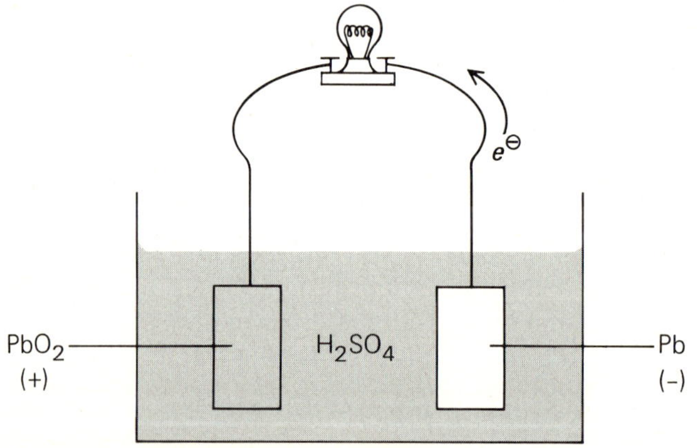

Figure 9.1
Voltaic cell.

We have seen that for the reaction to occur electrons must be transferred from the lead to the lead dioxide and hydrogen ions. In the cell, this can only happen if we connect the lead and lead dioxide by a wire. This electron flow is called an *electric current*. Thus the reaction, as it proceeds, gives out electrical energy and the bulb lights (in the bulb, electrical energy is converted into light energy and some heat).

ELECTRICAL ENERGY FROM LIVING ORGANS

We have seen how the chemical reactions in a voltaic cell produce an electric current. In a rather similar way certain chemical reactions in the human body produce electric currents. With suitable instruments these currents can be recorded. The pattern of the record indicates whether the organ is functioning normally or not. A well-known example of this is the electrocardiogram.

Electrocardiogram ECG

The electrocardiogram is the record of the electrical impulses produced by the chemical changes in contracting heart muscle. The instrument used for recording these impulses is the electrocardiograph. This is often portable so that it can be taken to the patient's bedside. Wires from the instrument are attached to each limb and recordings are taken from the following sites:

lead-I	right arm and left arm
lead-II	right arm and left leg
lead-III	left arm and left leg

In addition to these standard leads there are also chest leads and unipolar limb leads.

A contracting muscle is electrically negative compared to a resting muscle. So electrons tend to flow from contracting to resting muscles. In a normal heart there is an orderly progression of the muscular contraction and its associated electrical impulses over the walls of the atria and the ventricles (Fig. 9.2(a)). If there is damage to the heart muscle or to the special atrio-ventricular conducting system (see Fig. 9.5) this normal progression is disturbed. From the abnormal pattern produced by the electrical impulses on the ECG it is often possible to diagnose the abnormality present in the heart (Fig. 9.2(b)).

Electroencephalogram EEG

The electroencephalogram is the record of the electrical impulses produced by the chemical changes occurring in the living brain. The instrument used is the electroencephalograph. It is attached by numerous

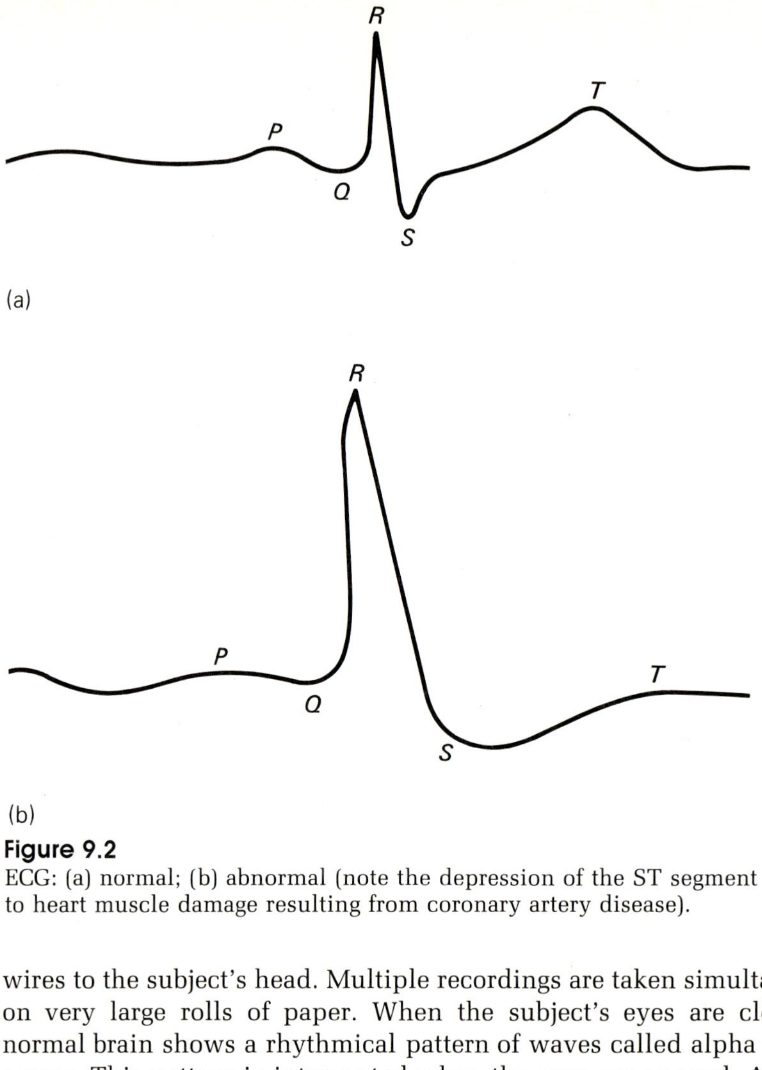

Figure 9.2
ECG: (a) normal; (b) abnormal (note the depression of the ST segment due to heart muscle damage resulting from coronary artery disease).

wires to the subject's head. Multiple recordings are taken simultaneously on very large rolls of paper. When the subject's eyes are closed the normal brain shows a rhythmical pattern of waves called alpha and beta waves. This pattern is interrupted when the eyes are opened. Abnormal patterns are seen from the brains of people who are subject to epilepsy or when a tumor is present. In view of the complexity of the equipment it is not portable. So the patient must be taken to the EEG room.

ELECTROLYTIC CELLS

Electrolytic cells are the reverse of batteries and enable electrical energy to be changed into chemical energy. This is done by passing an electric current through a compound when it is either molten (fused) or dissolved in water. This decomposes the compound and new substances are formed at the electrodes. Such a decomposition is called *electrolysis* and compounds capable of being decomposed in this way are known as electrolytes. Some compounds, for example, sodium chloride, are good conductors of electricity, in solution or when molten. We say they are *strong electrolytes*. Others, such as acetic acid, are poor conductors and are said to be *weak electrolytes*.

Electrolysis of sodium chloride

If an electric current is passed through molten sodium chloride (Fig. 9.3), chlorine gas bubbles off at the anode and molten sodium is formed at the cathode. How can we explain this decomposition? It is not too difficult. All we have to realize is that an electric current is a flow of electrons and that sodium chloride is an ionic compound consisting of sodium ions $Na^{\oplus}$ and chlorine ions $Cl^{\ominus}$.

The battery pumps electrons on to the cathode, making it negative so that it attracts the sodium ions $Na^{\oplus}$. When they reach the cathode they are reduced by the electrons to sodium atoms; thus, metallic sodium is formed.

$$Na^{\oplus} + e^{\ominus} = Na \qquad \text{(reduction)}$$

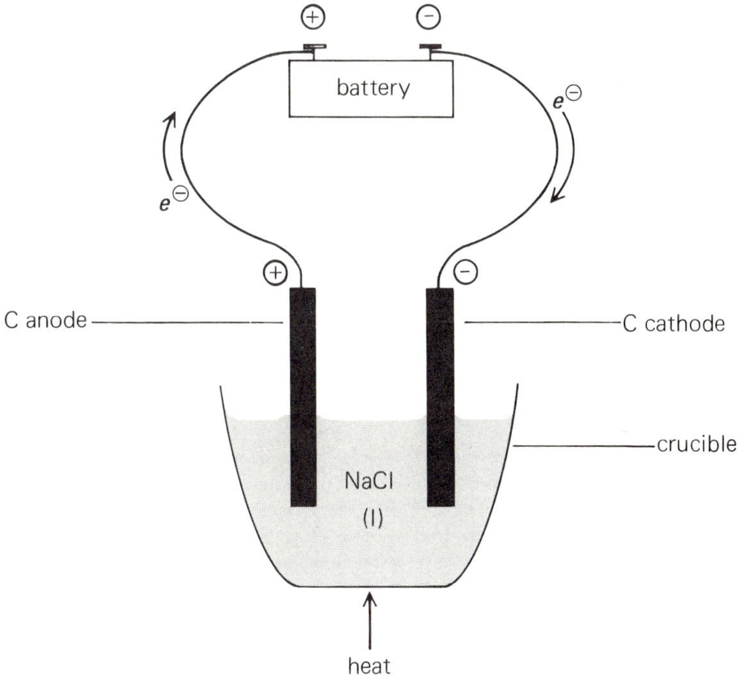

Figure 9.3
Electrolysis of sodium chloride.

The anode, being positive, attracts the chlorine ions $Cl^{\ominus}$, and these give up electrons to the anode.

$$2Cl^{\ominus} = Cl_2 + 2e^{\ominus} \qquad \text{(oxidation)}$$

The electrons then flow back to the battery, completing the circuit.

Sodium and chlorine are manufactured by this method; however, precautions are taken to prevent the molten sodium from combining with the chlorine to reform sodium chloride. It must also be kept from the air, otherwise it would combine with atmospheric oxygen to form sodium oxide.

Electrolysis of sodium chloride solution

If a *solution* of sodium chloride is used instead of *molten* sodium chloride, then electrolysis gives different products. Hydrogen gas bubbles off at the cathode and this is one way of producing hydrogen on a commercial scale. The product at the anode depends on the concentration of the salt solution. A strong solution gives chlorine and a weak one gives oxygen. Moderately concentrated solutions give both gases. How do we account for these different products? The big difference in this electrolysis is that water is present. This means that, at the cathode, sodium ions are not the only particles that can take up electrons; water molecules can as well, in the following manner.

$$\underset{(l)}{2H_2O} + 2e^{\ominus} = \underset{(g)}{H_2} + \underset{(aq)}{2OH^{\ominus}} \qquad \text{(reduction, occurring at the cathode)}$$

In fact, this is the reaction that occurs because it requires less energy than the reduction of sodium ions.

At the anode, if the concentration of chlorine ions is very low, water molecules are oxidized in preference to chlorine ions.

$$\underset{(l)}{2H_2O} = \underset{(aq)}{4H^{\oplus}} + \underset{(g)}{O_2} + 4e^{\ominus} \qquad \text{(oxidation, occurring at the anode)}$$

Thus, if a current is passed through a very dilute solution of sodium chloride, effectively what happens is that water is decomposed into hydrogen and oxygen.

The same result could be achieved by using pure water instead of sodium chloride solution, but in this case the electrolysis would take considerably longer. This is because the water molecules, being neutral, are not attracted to the positive and negative electrodes. We have to wait for their random motion to take them there. However, in a solution of sodium chloride, the sodium ions "carry" the water molecules to the

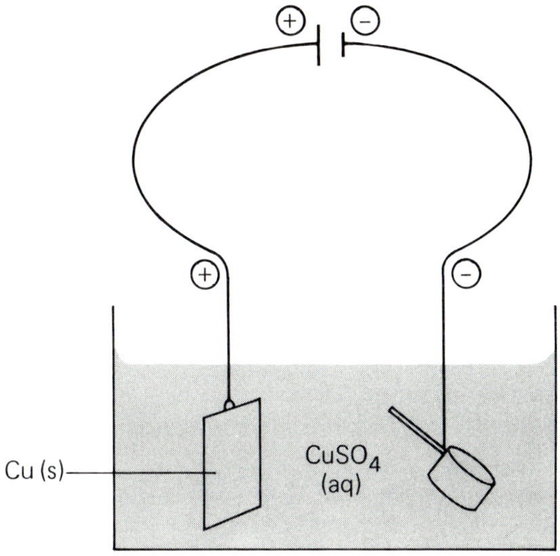

Figure 9.4
Copper electroplating.

cathode and the chlorine ions "carry" them to the anode. We shall find out in Chapter 12 how these ions are able to transport the water molecules.

Copper electroplating. The metal article, such as a saucepan, to be copper plated is made to be the cathode in a cell. This is shown schematically in Fig. 9.4.

The anode is made of copper and the electrolyte is copper sulfate. Electrolysis results in copper being evenly deposited on the saucepan (the cathode) and the copper anode dissolving.

The reduction at the cathode is

$$\underset{\substack{\text{(aq)}\\ \text{FROM COPPER}\\ \text{SULFATE}}}{Cu^{2\oplus}} + 2e^{\ominus} = \underset{\text{(s)}}{Cu}$$

and the oxidation at the anode is

$$\underset{\substack{\text{(s)}\\ \text{FROM ANODE}}}{Cu} = \underset{\text{(aq)}}{Cu^{2\oplus}} + 2e^{\ominus}$$

For every copper ion $Cu^{2\oplus}$ reduced at the cathode, a copper ion is produced at the anode. The concentration of the copper sulfate solution therefore remains constant.

EFFECT OF ELECTRIC CURRENT ON LIVING TISSUE

Too strong an electric current causes electrocution, as in the electric chair used for capital punishment, which results in the heart being stopped. If a relatively small electric current is applied to the nerve supplying a muscle it causes the muscle to contract. Muscle contraction is a complex chemical reaction in which physical energy is released. This reaction can also be triggered off by direct application of the electric current to the muscle. The nerve and muscle response to electrical stimulation is sometimes used clinically to determine the extent of damage in a case of paralysis. It is also used to prevent muscles from wasting while the nerve to the muscles is regenerating after an injury.

A fairly recent development of tissue stimulation by electric current is the artificial pacemaker which is surgically implanted to make the heart beat regularly when the normal mechanism has failed, as in com-

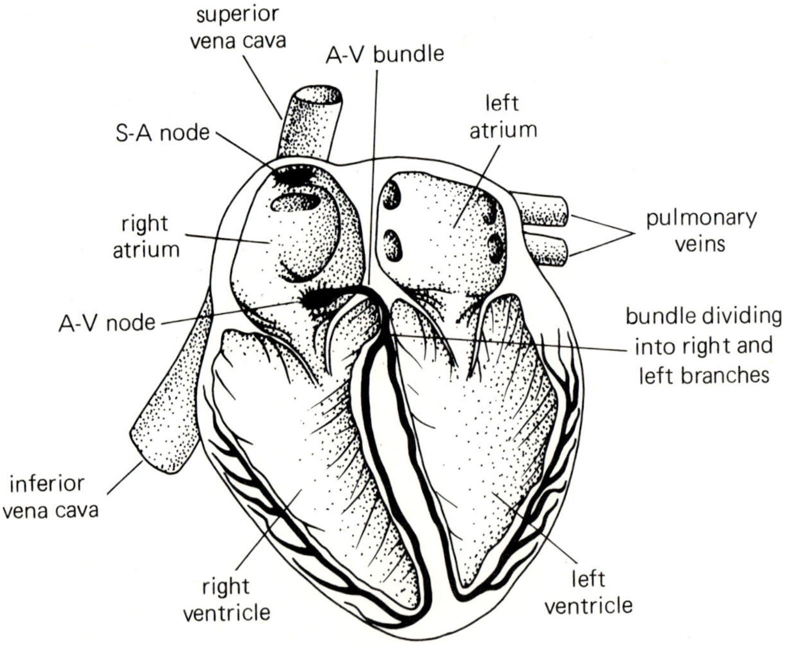

Figure 9.5
Heart, showing pacemaker, S-A node and conducting system, A-V node and bundle. The impulse originates from the S-A, Sino-Auricular, node, spreads over the walls of the atria and reaches the A-V, Atrio-Ventricular node. Then, it is conducted to the ventricles through the A-V bundle which divides into right and left branches to supply the two ventricles.

plete heart block in which the impulse fails to pass from atria to ventricles. Electrical stimulation is also being used to overcome "foot drop" and other forms of muscular paralysis. Another development is the *defibrillator* used in appropriate cases of heart attack to stop ventricular fibrillation, a rapid uncoordinated and ineffectual contraction of the ventricular muscle. Stopping this irregular contraction allows the normal rhythmical contraction to take place. This rhythm normally emanates from a pacemaker of specialised tissue in the right atrium called the sino-auricular, S-A node (Fig. 9.5). The production of the impulse by the pacemaker and the spread of the impulse, producing the orderly wave of muscular contraction involving first the atria and then the ventricles, are all electrochemical phenomena.

10

Oxygen

Our lives depend on oxygen. Cut off our oxygen, and within minutes we die. Brain cells die within 90 seconds of oxygen deprivation. However, breathing can stop for several minutes without fatal effects. This is because the air spaces in our lungs and the oxyhemoglobin in our blood provide a short-term oxygen reservoir. Since the reservoir is so quickly used up, it is vital for artificial respiration to be started at the earliest possible moment in respiratory paralysis due to electric shock or in cases of near drowning. Mouth to mouth respiration, the "kiss of life," has proved to be by far the most effective method.

If the heart stops even the short-term oxygen reservoir in lungs and blood is unavailable. There is no blood flow to the brain. So resuscitation has to be started within about a minute. Recently a number of people have had their hearts successfully restarted by cardiac massage or by electrical stimulation.

ATOMIC STRUCTURE

The atomic number of oxygen is 8 and its electron configuration is 2, 6. This means that the first or K shell is completely filled. The second or L shell, having 6 electrons, is 2 electrons short of the stable octet which is the inert gas configuration present in neon. This 2 electron gap in its outer valence shell gives oxygen a valence of 2. Diagrammatically the oxygen atom may be represented as:

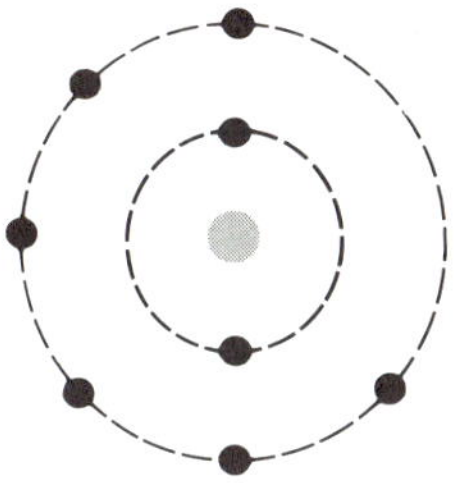

The valence shell of oxygen is relatively close to the nucleus, shielded from it only by the inner shell of 2 electrons. In addition, the positive charge on the nucleus produced by its 8 protons is relatively high. So if electrons of another atom approach the valence shell they are exposed to a strong nuclear attraction, making oxygen a ready electron acceptor. Oxygen is therefore highly reactive and a powerful oxidizing agent, especially at the time of release from combination with another atom when it is momentarily in the atomic state.

DISCOVERY AND APPLICATIONS

It is a strange fact that oxygen was discovered by two men quite independently within the space of two years, first in 1772 by Carle Scheele, a poor Swedish apothecary, and then in 1774 by Joseph Priestley, an English nonconformist preacher, who was the first to publish his discovery by several years. Priestley produced the gas by heating red oxide of mercury, using a magnifying ("burning") glass to focus the sun's rays. He described how "a mouse lived twice as long in the new air as in the same confined volume of common air." He breathed it himself and felt that it lightened his breathing. Although he recommended its use in medicine, even he might have been surprised at how much it is now used—in the operating theater on the anesthetic machine, in the resuscitation and intensive care wards, in the radiotherapy department, in the premature baby unit, in the hospital laboratory, and in the home care of respiratory and cardiac invalids. It is used in aviation medicine, being available on airliners in emergency if the pressurization fails, also in mountaineering, deep-sea diving, and in industry.

For medical use, oxygen is stored at a very high pressure, 120 atmospheres, in steel cylinders. These cylinders should not be exposed to mechanical shocks or to heat. To release the oxygen in a steady stream and at a safe pressure, a reducing valve must be used. Only a specially made lubricant may be used to grease this reducing valve.

OCCURRENCE

Oxygen composes 21% of the earth's atmosphere by volume (23% by weight). In combined form it is present in water, humans, animals, plants, and most rocks. The earth's crust is nearly 50% oxygen, considerably more than any other element.

PREPARATION

The preparation of oxygen provides a good example of the way a chemical reaction can be speeded up by a catalyst. Oxygen can be produced simply by heating potassium chlorate.

$$2KClO_3 \;=\; 2KCl \;+\; 3O_2$$

$2KClO_3$	$2KCl$	$3O_2$
(s)	(s)	(g)
POTASSIUM CHLORATE	POTASSIUM CHLORIDE	OXYGEN

This can bc done by placing the potassium chlorate crystals in a test tube and strongly heating it over a Bunsen burner. The crystals crackle and then melt. Only when the temperature is almost high enough to melt the tube is oxygen slowly produced. But if, before heating, a small amount (1 part in 5) of manganese dioxide[1] is added, oxygen is produced rapidly, even before the potassium chlorate melts.

The manganese dioxide is a *catalyst:* It speeds up the reaction but is not itself used up in the process. It can be compared to oil lubricating the mechanism of the reaction and allowing it to proceed with much less effort. In this example of *catalysis* the manganese dioxide allows the oxygen atoms to be pulled out of the potassium chlorate molecule without using so much energy in the form of heat. Catalysis is very important in chemistry. Some reactions do not take place at all without a catalyst. Many chemical reactions in the human body depend on catalysts made of protein, called enzymes.

Oxygen from water

We saw in Chapter 9 how water molecules can be electrolyzed, or split into their constituent elements by the action of an electric current.

$$\underset{(l)}{2H_2O} \longrightarrow \underset{(g)}{2H_2} + \underset{(g)}{O_2}$$

For the electrolysis of water the type of apparatus illustrated in Fig. 10.1 is suitable.

[1] Sometimes manganese dioxide contains impurities which cause the mixture with potassium chlorate to explode when heated.

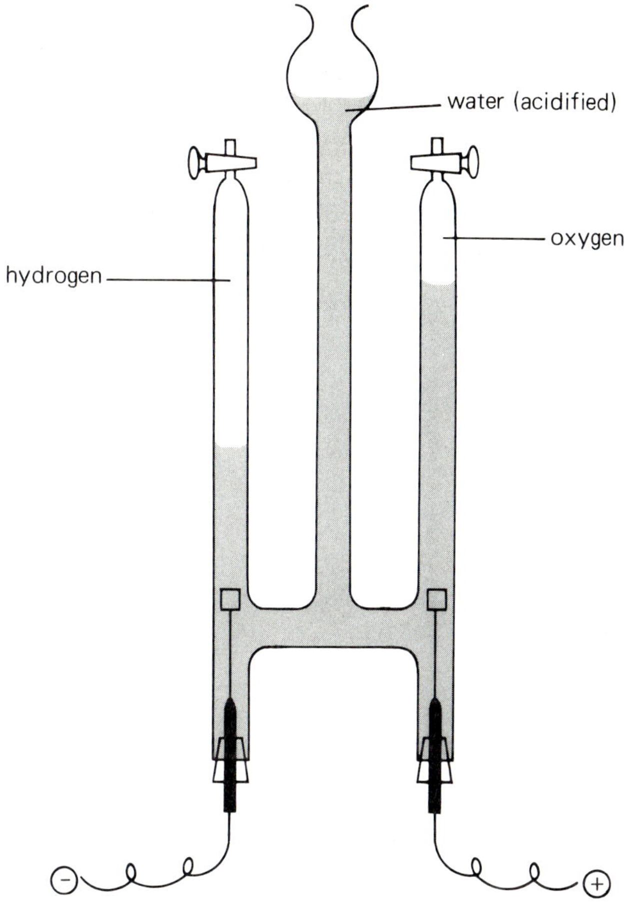

Figure 10.1
Electrolysis of water.

Oxygen comes off at the positive electrode (anode) and hydrogen at the negative electrode (cathode). The electrodes are made of platinum which does not corrode. The volume of hydrogen produced is twice that of oxygen and we see from the equation that the electrolysis gives twice as many hydrogen molecules as oxygen molecules. This illustrates a fundamental law of chemistry known as Avogadro's law, which states that equal volumes of gases contain the same number of molecules, provided that they are at the same temperature and pressure.

Oxygen from air

It is from the air that the oxygen supplied to hospitals and to industry is obtained. By a cooling process, similar to that used in a domestic refrigerator, the air is liquefied. From this liquid air, nitrogen boils off at

−196°C, leaving behind liquid oxygen which does not boil until the temperature reaches −183°C.

PROPERTIES

Oxygen is a colorless, odorless and tasteless gas. It has a very strong attraction for electrons and so it is termed very *electronegative.* So when oxygen combines with other elements it oxidizes them because it gains their electrons wholly or partly. An exception to this is its reaction with fluorine, the most electronegative of all elements. In this reaction fluorine oxide, F_2O is formed and the fluorine partially gains electrons from oxygen; so fluorine oxidizes oxygen (and oxygen reduces fluorine).

The most important chemical property of oxygen is its ability to support combustion. If a smoldering taper is plunged into a tube of oxygen it bursts into flame. If a patient on oxygen were to light a cigarette he or she could readily set himself or herself on fire. Fire is one of the major hazards of the oxygen tent. Three American astronauts perished by fire in an oxygen atmosphere while testing their space vehicle.

The process of combustion is simply the combination of the burning substance with oxygen. Coal for example is very largely carbon. When carbon burns it combines with the oxygen in the air to form carbon dioxide.

$$C + O_2 = CO_2$$

(s)	(g)	(g)
CARBON	OXYGEN	CARBON DIOXIDE

This reaction gives out heat so it is exothermic. As a result of giving out heat the temperature is raised. As the temperature rises the reaction takes place more rapidly. So heat is given out more rapidly. Thus, there is a vicious circle which can progress to white heat if there is a sufficient supply of combustible material and oxygen. In the domestic heating boiler the temperature is controlled by the air inlet valve. If this is opened wide a low fire soon bursts into life and the temperature rises.

When magnesium is ignited it burns with a blinding white flame and is oxidized to magnesium oxide.

$$2Mg + O_2 = 2MgO$$

(s)	(g)	(s)
MAGNESIUM	OXYGEN	MAGNESIUM OXIDE

In the process each magnesium atom loses two electrons to an oxygen atom.

$$Mg\,{}^{\times}_{\times} + \ddot{\underset{\cdot\cdot}{O}}: \longrightarrow Mg^{2\oplus}\quad {}^{\times}_{\times}\ddot{\underset{\cdot\cdot}{O}}:^{2\ominus}$$

When iron rusts it combines with oxygen and the rust is iron oxide. Noble metals such as platinum and gold are resistant to oxidation by the atmosphere. Base metals such as sodium and potassium combine readily with oxygen, as can be shown by using the apparatus illustrated in Fig. 10.2. Their oxides combine with water to form alkalis.

Some nonmetals such as sulfur, phosphorus, and carbon are also readily oxidized, but their oxides combine with water to form acids. From this property oxygen gained its name (Gr. *oxy*, acid, + Gr. *gen*, producer). Sulfur burns in oxygen with a blue flame. It can be demonstrated by placing a piece of roll sulfur in a deflagrating spoon, igniting it, putting it in a gas jar filled with oxygen (Fig. 10.2), and keeping the lid closed to contain the choking fumes.

The main product of combustion is sulfur dioxide.

S	+	O_2	=	SO_2
(s)		(g)		(g)
SULFUR		OXYGEN		SULFUR DIOXIDE

A little sulfur trioxide is also produced.

$2S$	+	$3O_2$	=	$2SO_3$
(s)		(g)		(g)
				SULFUR TRIOXIDE

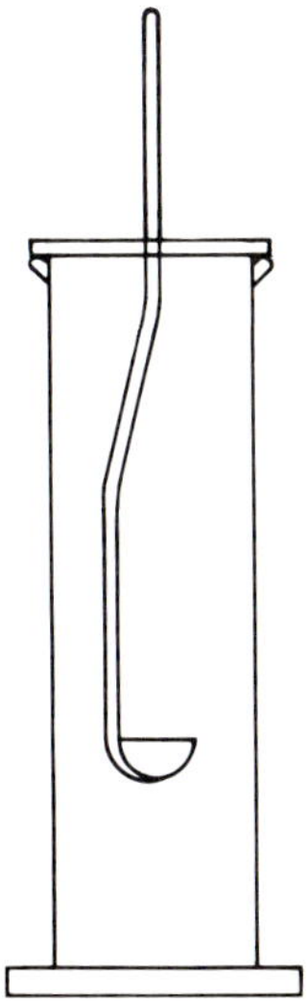

Figure 10.2
Gas jar and deflagrating spoon.

If water is introduced into the jar, both oxides dissolve in it to form acids.

$$\underset{\text{SULFUR DIOXIDE}}{\underset{(g)}{SO_2}} + \underset{\text{WATER}}{\underset{(l)}{H_2O}} = \underset{\text{SULFUR\underline{OUS} ACID}}{\underset{(aq)}{H_2SO_3}}$$

$$\underset{\text{SULFUR TRIOXIDE}}{\underset{(g)}{SO_3}} + \underset{\text{WATER}}{\underset{(l)}{H_2O}} = \underset{\text{SULFUR\underline{IC} ACID}}{\underset{(aq)}{H_2SO_4}}$$

The presence of acid may be shown by adding blue litmus, which turns red. Most of the acid is sulfurous acid with only a trace of sulfuric acid.

OXIDATION OF BACTERIA

Some antiseptics destroy bacteria by oxidation. Hydrogen peroxide is a well-known example. It is relatively unstable, readily losing an oxygen atom to any suitable nearby substance such as the wall of a bacterium and oxidizing it. So it produces its antiseptic action by virtually burning up the bacteria. Hydrogen peroxide is broken down on heating into water and molecular oxygen.

$$\underset{\text{HYDROGEN PEROXIDE}}{\underset{(l)}{2H_2O_2}} \longrightarrow \underset{\text{WATER}}{\underset{(l)}{2H_2O}} + \underset{\text{MOLECULAR OXYGEN}}{\underset{(g)}{O_2}}$$

A "20 volume" solution of hydrogen peroxide is one of such a concentration that 1 cm^3 of it, when heated, will give off 20 cm^3 of oxygen, measured at S.T.P. (S.T.P. stands for standard temperature and pressure which are 0°C and 760 mm of mercury pressure).

11

Hydrogen

Hydrogen is the lightest of all elements and has the smallest atoms. Thermonuclear reaction of hydrogen in the sun (see Chapter 32) provides the energy on which life on earth depends. Chemically, hydrogen is highly reactive; its reaction with oxygen provides the essential source of energy in our bodies.

ATOMIC STRUCTURE

Hydrogen has the simplest of all atoms. Each atom consists of a single proton around which orbits a single electron.

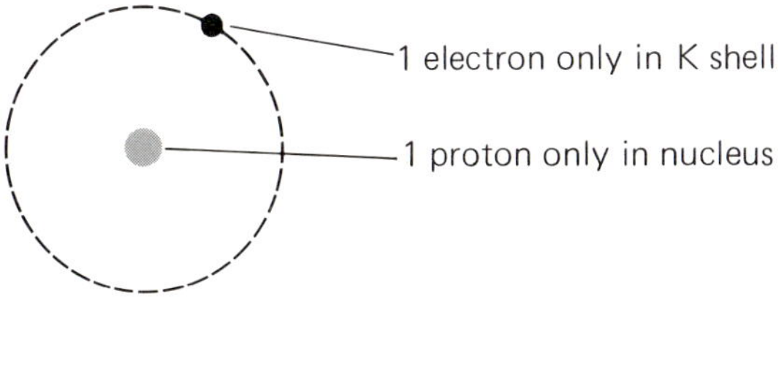

Unlike the atoms of all other elements the nucleus of hydrogen contains no neutrons. The hydrogen ion $H^{\oplus}$ is a hydrogen atom from which the single electron has been removed. It thus consists of only a positively charged proton.

ISOTOPES OF HYDROGEN

Its isotopes resemble hydrogen in that their atoms contain one proton and one electron, but in addition they also contain 1 or 2 neutrons in the nucleus

hydrogen = 1 electron + 1 proton

deuterium = 1 electron + 1 proton + 1 neutron

tritium = 1 electron + 1 proton + 2 neutrons

They may be represented diagrammatically as shown in Fig. 11.1.

Deuterium, D

Deuterium is twice as heavy as hydrogen; its oxide, D_2O is very similar to water, H_2O and is called "heavy water." It is demonstrably denser than ordinary water and has freezing and boiling points which are slightly higher. It occurs naturally, ordinary water containing a small

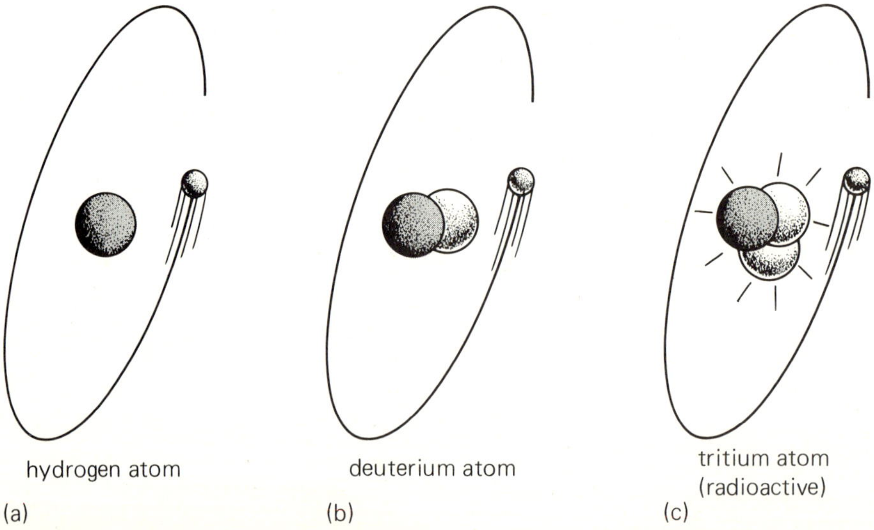

Figure 11.1

proportion (1 in 6000) of deuterium oxide. The chemical properties of deuterium are very similar to hydrogen but its reactions are somewhat slower.

Tritium, T

Tritium is three times as heavy as hydrogen. Unlike deuterium it does not occur naturally. It is made by bombarding deuterium atoms by deuterium ions in a cyclotron. Tritium has proved very useful in studies on living cells. For example, it can be introduced into the molecule of thymidine to form tritiated thymidine. This is taken up by actively dividing cells in which it can be demonstrated by its radioactivity, using autoradiographs (the radioactivity produces dark spots on photographic film). In this way it is easy to see which are the rapidly dividing cells. This has been of considerable value in cancer research.

DISCOVERY

Hydrogen was first described to the Royal Society, London, by Henry Cavendish in 1766 as "inflammable air." Several years later he showed that a mixture of "inflammable air" and oxygen could be exploded to form water. So he called his newly discovered gas hydrogen which means water producer (Gr. *hydro*, water, + Gr. *gen*, producer). It is probable, however, that hydrogen had been prepared about a century earlier by Robert Boyle, from iron filings and hydrochloric acid.

OCCURRENCE

More than 90% of the universe consists of hydrogen, either as an element or in compounds. The sun consists mainly of hydrogen and every second 4 million tons of it are converted directly into energy. A tiny fraction of this sunlight is trapped by the chlorophyll in green plants on earth, providing the energy for all the other living creatures including ourselves, which feed directly or indirectly on green plants.

Not only the sun but also all the other billions of stars are composed mainly of hydrogen. So are the huge interstellar gas clouds. On earth, however, hydrogen is only the ninth most common element by weight after oxygen, silicon, aluminium, iron, calcium, sodium, potassium, and magnesium.

Hydrogen is continually escaping from the earth's atmosphere, of which it forms only one part in a million by volume (much less by weight). So free hydrogen is rare except where it has been trapped underground, as in North Sea gas, or released from compounds by man's efforts, as in coal gas (which also contains the highly poisonous carbon monoxide).

Hydrogen is most commonly found on earth in the form of its oxide, water, H_2O which covers much of the earth's surface. Compounds of hydrogen with carbon alone are called hydrocarbons. They are found in coal and petroleum which are derived from plants and creatures of prehistoric times. Examples of hydrocarbons are paraffin, candle grease, and petrol. They must be distinguished from carbohydrates. These are the sugars and starches, which contain oxygen as well as carbon and hydrogen.

Acids are a very important group of substances containing hydrogen. It is hydrogen in the form of the hydrogen ion, $H^{\oplus}$ which gives acids their bite. The acidity of a solution depends on the concentration of hydrogen ions. We saw in Chapter 5 how these ions each combine with a molecule of water to form a hydroxonium ion $H_3O^{\oplus}$.

PREPARATION

In the laboratory, hydrogen is usually prepared by the action of dilute sulfuric or hydrochloric acid on granulated zinc.

Zn	+	H_2SO_4	=	$ZnSO_4$	+	H_2
(s)		(aq)		(aq)		(g)
ZINC		SULFURIC ACID		ZINC SULFATE		HYDROGEN

A suitable apparatus is illustrated in Fig. 11.2, the acid being introduced through a dropping funnel into a flask containing zinc granules. The hydrogen gas is collected over water in an inverted jar or in a gas syringe. Note that until the air has been blown out of the apparatus the gas collected may be an explosive mixture of hydrogen and air. A small volume collected into a test tube should first be tested. When pure hydrogen is coming over it will burn without exploding.

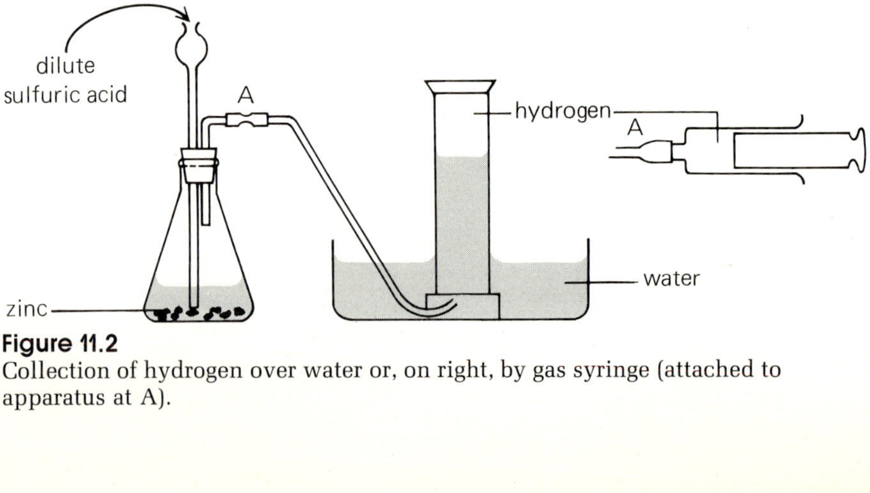

Figure 11.2
Collection of hydrogen over water or, on right, by gas syringe (attached to apparatus at A).

Hydrogen from water

By electrolysis. The passage of an electric current through water (slightly acidified to increase its electrical conductivity) causes it to dissociate into hydrogen and oxygen.

$$\underset{(l)}{2H_2O} = \underset{(g)}{2H_2} + \underset{(g)}{O_2}$$

Electrolysis is used commercially for the preparation of hydrogen from solutions of sodium hydroxide or sodium chloride.

By chemical methods. Several of the baser metals readily displace hydrogen from water. With sodium or potassium the reaction is violent; only a small piece, about the size of a pea, should be used. So much heat is produced in the reaction that the metal melts and shoots across the surface of the water owing to rapid evolution of hydrogen. In the case of potassium, the heat is sufficient to ignite the hydrogen.

White hot coke also displaces hydrogen from steam.

$$\underset{\substack{(s)\\ \text{COKE}}}{C} + \underset{\substack{(g)\\ \text{STEAM}}}{H_2O} = \underset{\substack{(g)\\ \text{CARBON}\\ \text{MONOXIDE}}}{CO} + \underset{\substack{(g)\\ \text{HYDROGEN}}}{H_2}$$

The mixture of carbon monoxide and hydrogen which is formed is known as *water gas*. It is a useful fuel and an industrial source of hydrogen. It is also used in the manufacture of methanol mentioned later in the chapter.

PROPERTIES

Hydrogen gas is the lightest of all elements, being fourteen times lighter than air. Its atomic number is one and its atomic weight is also approximately one, virtually all its mass being related to the single proton. It was at one time frequently used for filling balloons and airships but was highly dangerous because of its flammability. Helium is now used for airships instead.

Free hydrogen cannot be detected by our senses as it is colorless, odorless and tasteless. It liquefies at $-253°C$ to form a pale blue liquid, the lightest liquid known, being fourteen times lighter than water. When it is cooled only 8° below its boiling point it freezes into solid hydrogen.

Chemically, hydrogen is very reactive. It has a strong attraction for oxygen and chlorine. A jet of hydrogen will burn readily in either gas forming water with oxygen and hydrogen chloride with chlorine.

$$\underset{(g)}{2H_2} + \underset{(g)}{O_2} = \underset{(l)}{2H_2O}$$

$$\underset{(g)}{H_2} + \underset{(g)}{Cl_2} = \underset{(g)}{2HCl}$$

If hydrogen is mixed with oxygen or chlorine and the mixture ignited, the same reactions occur as before, but this time with explosive violence.

Hydrogen combines much less readily with nitrogen, neither exploding nor burning in it. The reaction is a reversible one.

$$\underset{(g)}{N_2} + \underset{(g)}{3H_2} \rightleftharpoons \underset{(g)}{2NH_3}$$

Under ordinary atmospheric conditions the amount of ammonia produced is negligible. But by increasing the pressure to 1000 atmospheres the yield can be increased to about 60%. The speed of the reaction is increased by raising the temperature to 500°C and using catalysts of iron and molybdenum. The ammonia produced, as in the Haber process, is of considerable importance industrially for making fertilizers such as ammonium sulfate, $(NH_4)_2SO_4$ and in the production of nitric acid, HNO_3.

Hydrogen reacts with certain vegetable oils to produce edible fats, such as margarine. This is known as the "hardening" of oils and requires a catalyst. Hydrogen is also used in the manufacture of methanol (methyl alcohol) from water gas.

$$\underbrace{\underset{(g)}{CO} + \underset{(g)}{H_2}}_{\text{WATER GAS}} + \underset{(g)}{H_2} = \underset{\substack{(l)\\ \text{METHANOL}}}{CH_3OH}$$

Since hydrogen combines with oxygen so readily it can be used to free metals from their oxides. Molybdenum and tungsten are manufactured in this way.

$$\underset{\substack{(s)\\ \text{MOLYBDENUM}\\ \text{OXIDE}}}{MoO_3} + \underset{(g)}{3H_2} = \underset{\substack{(s)\\ \text{MOLYBDENUM}}}{Mo} + \underset{(l)}{3H_2O}$$

In all of the above reactions hydrogen is a reducing agent because it partially loses its electrons to other elements, namely oxygen, chlorine, nitrogen, and carbon. It can also act as an oxidizing agent and gain electrons as in the formation of sodium and potassium hydrides.

$$\underset{\text{SODIUM}}{\underset{(s)}{2Na}} + \underset{\text{HYDROGEN}}{\underset{(g)}{H_2}} = \underset{\text{SODIUM HYDRIDE}}{\underset{(s)}{2NaH}}$$

This occurs when sodium is heated to 400°C in a stream of hydrogen. Sodium hydride is an ionic compound $Na^{\oplus} H^{\ominus}$. It is a white crystalline substance and is very unstable, reacting immediately with water to release hydrogen.

$$\underset{(l)}{H_2O} + \underset{(s)}{NaH} = \underset{(aq)}{NaOH} + \underset{(g)}{H_2}$$

HYDROGEN IN THE BODY CHEMISTRY

Hydrogen compounds are widely distributed in cells and body fluids. The oxide, water, is so important that we treat it separately in Chapter 12. The part played by hydrogen ions in body acidity will be considered in Chapter 13. In the many organic compounds which provide the body's building bricks, communication systems, fuel supplies, and so forth, attention is so frequently focused on the carbon atoms that the more numerous hydrogen atoms tend to be overlooked. Sometimes, however, these hydrogen atoms make a major contribution, particularly in energy production. We have seen how the combination of hydrogen with oxygen to form water provides so much energy that if the proportions are approximately right, a mixture of these gases is explosive

$$2H_2 + O_2 = 2H_2O + \text{energy}$$

The energy produced is very large, 239 kilojoules for each mole of H_2O formed. This reaction, suitably controlled, is the essential source of the body's energy. As we have already seen, it occurs inside cells in the mitochondria. The oxygen on the left-hand side of the equation is supplied in aqueous solution from atmospheric oxygen via the lungs and red cell hemoglobin. The hydrogen is provided not as the free element but attached to a complex molecule NAD (*n*icotinamide-*a*denine *d*inucleotide) or NADP (*n*icotinamide-*a*denine *d*inucleotide *p*hosphate), which combines reversibly with hydrogen to form NADH or NADPH.

In this combined form the hydrogen does not react with the oxygen directly but through a number of intermediate steps known as the respiratory chain. These steps involve a series of enzymes such as flavoproteins and cytochromes situated in the mitochondria. The energy is liberated not as heat but as chemical energy stored in the energy-rich compound, ATP (*a*denosine *trip*hosphate). In this form, it is readily available to be converted into energy, such as the physical energy of muscular contraction.

12

Water

Water is one of the commonest chemical substances in the world. It is essential for all forms of life on this planet and comprises about 70% of our body weight. Approximately 20% of our weight is the water which is outside our cells, called *extracellular water*. Part of this is in our blood but most of it is in the fluid which surrounds our tissue cells, known as the *tissue fluid*. About 50% of our body weight is water situated inside our cells and called *intracellular water*.

We are continually losing water from our lungs in our breath, from our skin as sweat which may be imperceptible, from our kidneys in urine and from our alimentary tract in feces. Normally our water loss is counterbalanced by our water intake in the form of drink and food. Even solid food contains much water, and succulent fruit contains well over 90% water.

In some diseases there is excessive water loss, for example in the urine of a diabetic or in the diarrhea which accompanies food poisoning, dysentery and cholera, also typhoid after the initial stage of constipation. In all these conditions the body becomes depleted of water and the patient is dehydrated. In dehydration it is the extracellular water which is lost. Loss of intracellular water is incompatible with life. Replacement of lost water, particularly in infants who have so little reserve, may well be a life-saving measure.

NATURAL WATER

On this planet there is water everywhere. It occurs in the air as water vapor (rain) and snow; in the oceans and inland waters as liquid water, icebergs and glaciers; and in many rocks and minerals of the earth's crust as water of crystallization (see page 77). In its liquid state the purest form of natural water is the rain water of country districts. Even this is not absolutely pure. It contains dissolved oxygen, nitrogen, and carbon dioxide from the atmosphere. During thunderstorms it dissolves the nitric acid produced by the electrical discharges. In towns, it also contains dust, soot, and oxides of sulfur that are produced when coal is burnt.

The carbon dioxide dissolved in rain water makes it a very weak acid

$$\underset{\substack{(l)\\ \text{WATER}}}{H_2O} + \underset{\substack{(g)\\ \text{CARBON DIOXIDE}}}{CO_2} = \underset{\substack{(aq)\\ \text{CARBONIC ACID}}}{H_2CO_3}$$

After the rain water falls on to the ground it passes into the soil and eventually gathers into streams and rivers. From the soil and rocks it dissolves substances which are soluble either in pure water or in a dilute solution of acid. So the impurities present in water are determined by the types of soluble substances in the soil and rocks over which it has passed. Commonly, these are salts such as chlorides, carbonates, bicarbonates, and sulfates of sodium, calcium, magnesium, and iron. The nature of salts will be discussed in the next chapter.

SOFT AND HARD WATER

Soft water lathers readily with soap. Rain water is an example. With hard water considerably more soap is needed before a lather is obtained. An example is the hard water from a limestone district containing calcium bicarbonate. The salts of calcium, magnesium and iron cause water to be hard because they react with soap to form an insoluble precipitate. Soap is sodium stearate, represented for simplicity as Na"St".

$$\underset{\substack{(s)\\ \text{SODIUM STEARATE}}}{2Na``St"} + \underset{\substack{(aq)\\ \text{CALCIUM BICARBONATE}}}{Ca(HCO_3)_2} = \underset{\substack{(s)\\ \text{CALCIUM STEARATE (INSOLUBLE PRECIPITATE)}}}{Ca``St"_2} + \underset{\substack{(aq)\\ \text{SODIUM BICARBONATE}}}{2NaHCO_3}$$

It can be seen that the calcium from the bicarbonate replaces the sodium in soap to form calcium stearate which precipitates and forms a scum. Only when enough soap has been used to remove all the calcium as insoluble calcium stearate is it possible to start obtaining a lather. Sometimes the hardness of water can be removed by boiling it. This is called temporary hardness. If boiling does not remove it, it is called permanent hardness.

Temporary hardness is due to the presence of calcium or magnesium bicarbonate in the water. If the water is boiled the bicarbonate breaks down.

$$\begin{array}{ccccccc} Ca(HCO_3)_2 & = & CaCO_3 & + & H_2O & + & CO_2 \\ (aq) & & (s) & & (l) & & (g) \\ \text{CALCIUM BICARBONATE} & & \text{CALCIUM CARBONATE} & & \text{WATER} & & \text{CARBON DIOXIDE} \end{array}$$

Carbon dioxide is given off as gas and the calcium carbonate precipitates, as "fur" in a kettle. When this occurs the water becomes soft and lathers easily.

Permanent hardness is due to the presence of calcium or magnesium sulfate in the water. It is resistant to boiling.

Both temporary and permanent hardness may be removed by passing the water over a substance called sodium zeolite used in water softeners. This exchanges sodium ions for calcium or magnesium ions.

$$\begin{array}{ccccccc} 2NaZ & + & Ca^{2\oplus} & \rightleftharpoons & CaZ_2 & + & 2Na^{\oplus} \\ (s) & & (aq) & & (s) & & (aq) \\ \text{SODIUM ZEOLITE} & & \text{CALCIUM ION} & & \text{CALCIUM ZEOLITE} & & \text{SODIUM IONS} \end{array}$$

So the calcium and magnesium ions in the water are replaced by sodium ions and the water is made soft. The above reaction is reversible, as denoted by the "$\leftrightharpoons$" sign: the calcium or magnesium zeolite can be converted back to sodium zeolite, the active softening ingredient, simply by flushing it through with a solution of sodium chloride (common salt, NaCl) which provides sodium ions.

$$\begin{array}{ccccccc} CaZ_2 & + & 2Na^{\oplus} & \longrightarrow & 2NaZ & + & Ca^{2\oplus} \\ (s) & & (aq) & & (s) & & (aq) \\ \text{CALCIUM ZEOLITE} & & \text{SODIUM IONS (FROM BRINE)} & & \text{SODIUM ZEOLITE} & & \text{CALCIUM ION} \end{array}$$

Thus exhausted zeolite can be recharged in brine (a NaCl solution) overnight, ready for use again in the morning. Sodium zeolite, represented for simplicity as NaZ, is sodium aluminum silicate ($NaAlSi_2O_6$).

The advantages of soft water for washing are that less soap is needed, a better lather is obtained and no scum (calcium or magnesium stearate) is produced. Because less soap or detergent is needed, the skin of the hands suffers less from washing up or scrubbing up.

For laundries, soft water is much more economical in soap or detergent than hard water. For heating systems soft water leaves no "scale" of calcium or magnesium carbonates in the pipes, radiators or boilers which, if hard water is used, renders them progressively more inefficient, for the scale is a poor conductor of heat and the flow of water is impaired.

DRINKING WATER

There is no need to soften drinking water. In fact it could be harmful, for in areas where the water is soft a higher incidence of heart disease has been noted. The small amounts of salts present in hard water are probably beneficial and may give the water a pleasant flavor. Very soft water can dissolve lead from lead pipes and contribute to slowly cumulative lead poisoning.

Water containing nitrates is generally unsuitable for drinking. It is likely to be contaminated by sewage or decaying animal matter. Also, nitrates are reduced to nitrites by bacteria in the gut. These react with hemoglobin forming methemoglobin which is unable to transport oxygen. Drinking water should be free from bacteria. They may be removed by filtration through sand or killed, for instance, by chlorine. The most common way in which water is made safe for drinking is by chlorination. For small quantities of water, tincture of iodine may be used: 3 drops to 1 pint of water for 1 hour. The iodine can then be removed by 1 to 2 drops of sodium thiosulfate solution, to make the water palatable.

DISTILLED WATER

When water is boiled, salts or other dissolved solids are left behind and if the steam is condensed and collected (see Fig. 12.1) it provides water which is sufficiently pure for the majority of purposes where chemically pure water is required. *De-ionized* water is similar, the salts being removed by passing the water through an ion-exchange resin (see Fig. 12.2). Hydrogen ions, $H^{\oplus}$, in the resin are exchanged for positive ions, such as $Ca^{2\oplus}$, that are in the water. The resin also contains hydroxyl ions,

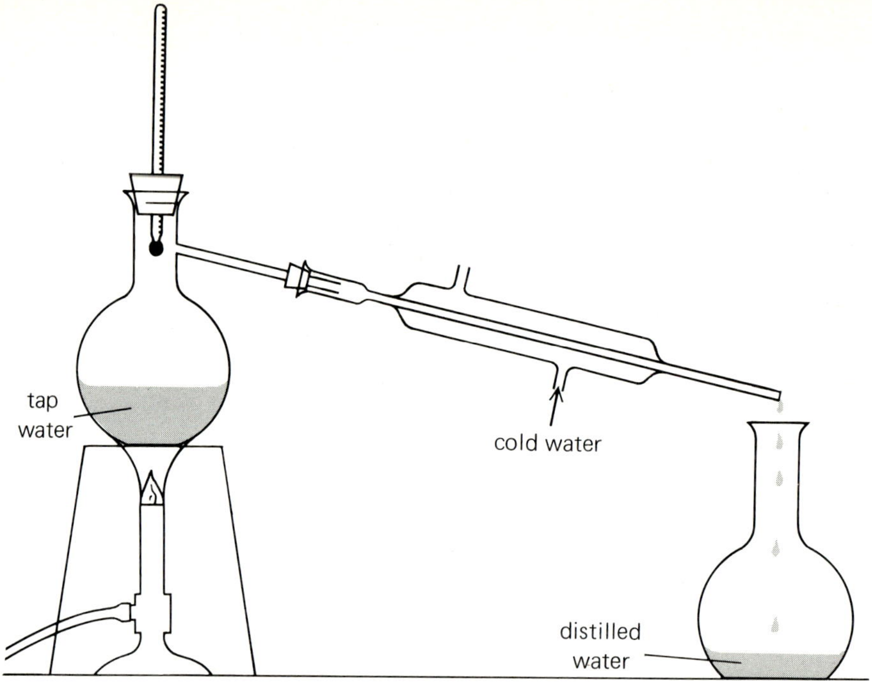

Figure 12.1
Distillation: the water is evaporated by heat and the vapor condensed by cooling.

$OH^{\ominus}$ and these are exchanged for negative ions, such as $Cl^{\ominus}$. The hydrogen and hydroxyl ions combine to form water.

$$\underset{(aq)}{H^{\oplus}} + \underset{(aq)}{OH^{\ominus}} = \underset{(l)}{H_2O}$$

Water may be further purified by repeated distillation. To obtain very pure water, distilled water is partially frozen and the ice crystals collected. Soluble impurities remain in the unfrozen water.

STERILE WATER

The concentration of chlorine (0.2 ppm) used to make water safe to drink is sufficient to kill off organisms of the coliform, typhoid or cholera type. Some spore-bearing organisms may, however, survive in tap water. These are harmless as far as drinking is concerned but the water cannot be considered as sterile for injection purposes. In addition the taps them-

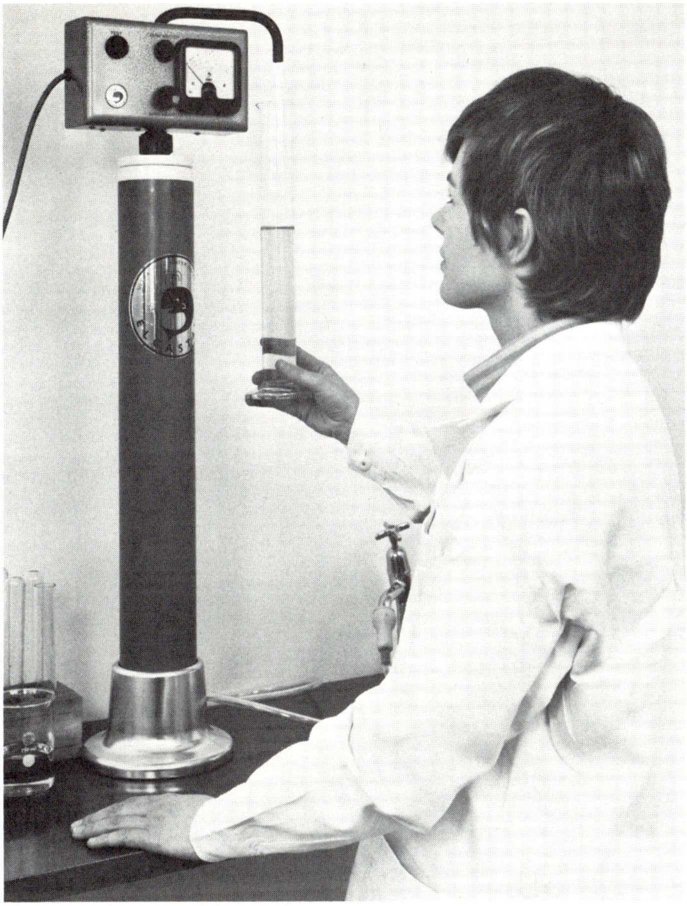

Figure 12.2
De-ionizer.

selves may harbor organisms, such as mycobacteria. When sterile water is required for injections or parenteral infusions it is usually heated in an autoclave (Fig. 12.3) at a pressure of 15 lb/in.^{2} (1.05 kg/cm^2) for 15 minutes (at 121°C). This is sufficient to kill all known organisms.

PROPERTIES OF WATER

Pure water is an odorless, tasteless liquid. Small amounts of water and ice appear colorless, but light which has penetrated through much water or ice is blue. A beautiful example of this is in the "Blue Grotto" on Capri Island. Daylight enters the cave through an underwater opening and the grotto is bathed in an eerie blue light.

The boiling point of water at 100°C and its freezing point at 0°C, at 1 atmosphere pressure, are used as the standards for calibrating thermometers. Water is an unusual liquid in having its maximum density at a tem-

Figure 12.3
Autoclave.

perature 4°C higher than its freezing point. At this temperature 1 cm^3 weighs 1 g. When water cools below 4°C it becomes less dense. And so when the temperature nears its freezing point the coldest water is on the surface. This is the reason why ice forms on the surface of a pond and not at its bottom, a matter of some importance for aquatic creatures and for us. For if ice formed at the bottom, melting would be a much slower process; much more ice in this world would never melt at all.

Water is a powerful solvent, particularly for acids, bases and salts (see Chapter 13). A number of salts will not form crystals unless they are combined with a definite proportion of water, known as *water of crystallization*, for example, magnesium sulfate crystals $MgSO_4 \cdot 7H_2O$ (Epsom salts) and nickel chloride crystals $NiCl_2 \cdot 6H_2O$. Many chemical reactions will not take place except in the presence of at least some water. The chemical reactions in the human body all take place completely immersed in a watery solution.

Meaning of formula H_2O

Do water's properties fit in with a formula of H_2O? It is frequently represented like this, but a molecule consisting of two atoms of hydrogen and one of oxygen has a molecular weight of only 18. A substance with such a light molecule should be a gas at room temperature. Hydrogen sulfide, H_2S has a molecular weight of 34, and it is certainly a gas, right down to temperatures well below the freezing point of water. How then can water, a liquid, have a formula of H_2O?

The explanation lies in the shape of the water molecule and in its ability to link with neighboring molecules.

Shape of a water molecule

When hydrogen combines with oxygen to form water, the bonds between the atoms are of covalent type.

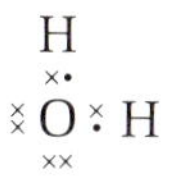

or

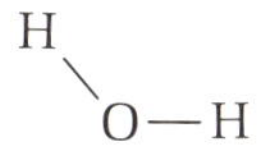

In the formula on the right the bonds are represented by lines. The angle between the two bonds is just over a right angle (104.5°). This odd shape provides part of the explanation of the unusual properties of water.

Linking of water molecules

In a water molecule the positive electrical charge on the nucleus of the oxygen atom is eight times greater than that on each of the hydrogen atoms. So the negatively charged electrons are attracted very slightly away from the hydrogen atoms towards the oxygen atoms. This produces a weak positive charge on the hydrogen atoms and a weak negative

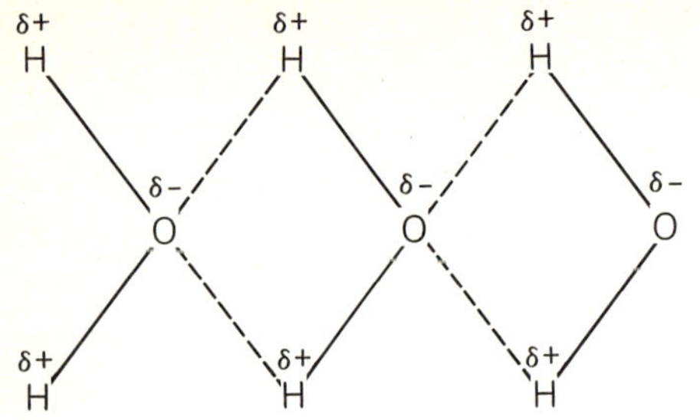

Figure 12.4
Hydrogen bonding between water molecules.

charge on the oxygen atom. We say that the water molecule is *polarized* since it contains positive and negative "poles."

As a result the negatively charged oxygen atom is attracted to the positively charged hydrogen atom of a neighboring molecule. This is an example of the hydrogen bond (Chapter 5). It links together groups of 3 or 4 water molecules to form one large molecular structure. In these groups the slightly positive hydrogen atoms of one H_2O molecule are attracted to the nonbonding electron-pairs in the oxygen atom of another H_2O molecule which gives a rather open structure to the grouping (see Fig. 12.4).

It is these large molecular structures which account for water being a liquid and not a gas at ordinary temperatures. As water is heated the energy imparted to the water molecules tends to disrupt the relatively weak hydrogen bonds, until at its boiling point water breaks down completely into individual molecules of H_2O to form gaseous steam.

The most densely packed molecular structure of water is at 4°C. Below this temperature, the groups of H_2O molecules grow bigger. Be-

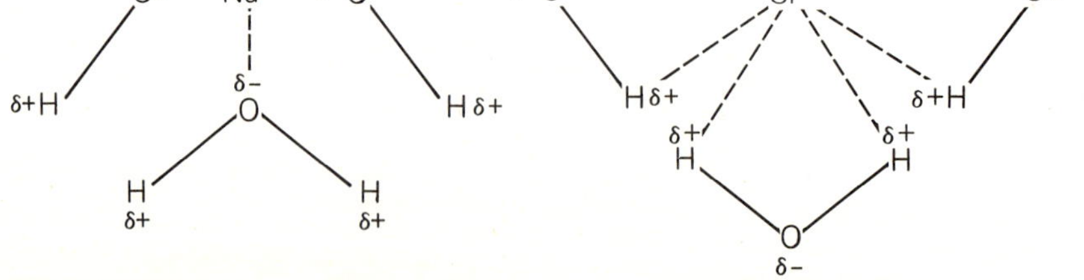

Figure 12.5
Sodium and chlorine ions surrounded by water molecules.

cause these groups have an open structure the density of water decreases as the groups grow. In ice all the H_2O molecules are hydrogen-bonded to form one gigantic grouping. Ice is therefore less dense than liquid water and so it floats.

Another important consequence of the polarization of water is that water molecules are weakly attracted to dissolved ions. You may recall from Chapter 9 that sodium and chlorine ions carry water molecules to the electrodes during the electrolysis of sodium chloride solution. They are able to do this because each sodium ion ($Na^{\oplus}$) attracts the slightly negative oxygen atoms of water molecules and each chlorine ion ($Cl^{\ominus}$), attracts the slightly positive hydrogen atoms. Thus the ions are surrounded by a "sphere" of water molecules (Fig. 12.5).

In Chapter 11 it was stated that a little acid accelerates the electrolysis of water. It does so for the same reason that sodium chloride does. The ions furnished by the acid ($H^{\oplus}$ and $SO_4^{2\ominus}$ ions in the case of sulfuric acid, H_2SO_4) carry the water molecules to the electrodes.

This attraction of water molecules for ions also explains why water is a good solvent for most ionic compounds. When a substance dissolves, its particles (ions or molecules) must mix thoroughly with the solvent molecules. If the particles are attracted to, and consequently surrounded by, molecules of solvent then thorough mixing is made possible.

13

Salts, Acids, and Bases

The salts dissolved in our body water are as essential to life as the water itself. The severe loss of fluid from the gut in cholera, from the kidneys in diabetes or from the skin in heat exhaustion also involves a severe loss of salts. In the treatment of these conditions it is vital to replace the salts as well as the water.

Chemically the salts dissolved in our blood and tissue fluid are very similar to those found in seawater. The concentration of the salts in the sea has been gradually increasing over the ages. In inland seas such as the Dead Sea and the Great Salt Lake in Utah, evaporation has resulted in the concentration of the salts being four to eight times that in the open ocean, so strong that fish cannot survive in it. In the open seas the concentration of salts has also been increasing, although more slowly.

Millions of years ago when the first living creatures appeared on earth the concentration of salts in the seas was much weaker than it is now. It was practically the same as that still found today in our blood and tissue fluid. So our cells are bathed by a salt solution much like that of the seas in prehistoric time when the first single-celled creatures (protozoa) were formed.

The salt present in greatest amount, both in our body fluid and in the sea, is sodium chloride NaCl or common salt. The chemical symbol for sodium Na, is derived from natrium, the medieval Latin name for sodium. There are also salts of potassium, K (L. *kalium*), calcium, Ca, and magnesium, Mg. To understand the chemistry of salts we must first discuss acids and bases.

ACIDS

We have all had some contact with acids, such as the sulfuric acid in car batteries or the hydrochloric acid from our stomachs. Characteristically, acids have a sour taste and tend to burn the skin. They can dissolve fabrics and have often been known to make holes in nylons. They can also corrode metal, as you may have noticed if you have had a leaking car battery. They react with carbonates and bicarbonates, giving off carbon dioxide, and they turn blue litmus red. These are typical acidic properties, but acids behave in this way only if they are dissolved in a suitable solvent, usually water. The reason for this is that the characteristic properties of acids are due to the proton (that is, the hydrogen ion $H^{\oplus}$) which is unable to exist on its own, under normal conditions. It has to attach itself to a suitable molecule, for example, H_2O. Thus, when the gas, hydrogen chloride, is dissolved in water it gives protons to the water molecules

$$HCl + H_2O = H_3O^{\oplus} + Cl^{\ominus}$$

The solution formed is known as hydrochloric acid. Sulfuric acid behaves in a similar fashion

$$H_2SO_4 + 2H_2O = 2H_3O^{\oplus} + SO_4^{2\ominus}$$

Acids are therefore givers of protons, that is *proton donors*. You can see that hydrochloric acid and sulfuric acid have an ion in common, namely $H_3O^{\oplus}$. This is a protonated water molecule which we met earlier in Chapter 5. You may remember that it is called the hydroxonium ion and that its structure is $H_2O \rightarrow H^{\oplus}$, a dative covalent bond linking the proton to the oxygen of the water molecule. It is the hydroxonium ion that confers on aqueous solutions of acids their typical properties.

Sulfuric, hydrochloric and nitric acids are strong proton donors and are therefore said to be strong acids. By contrast, acetic acid is a weak proton donor, that is, a weak acid. Only a small fraction of acetic acid molecules are able to protonate water molecules, whereas all the dissolved hydrogen chloride molecules are successful in this respect. This means that a solution of acetic acid contains a smaller concentration of hydroxonium ions than an equimolar solution (that is, one containing the same number of moles per liter) of hydrochloric acid. In other words, the acidic properties of the acetic acid solution will be less pronounced than those of the hydrochloric acid.

If hydrogen chloride is dissolved in a liquid that will not readily accept protons, for instance, glacial acetic acid (100% acetic acid), then the solution will contain only a small concentration of protonated solvent, because only a small proportion of hydrogen chloride molecules will be successful in forcing protons on to solvent molecules.

$$\underset{}{HCl} + \underset{\text{ACETIC ACID}}{C_2H_4O_2} \rightleftharpoons \underset{\text{PROTONATED ACETIC ACID}}{C_2H_5O_2^{\oplus}} + Cl^{\ominus}$$

In this case it is the $C_2H_5O_2^{\oplus}$ ion that confers acidic properties on the solution, and because their concentration is low, the acidic properties will not be as marked as they are in an aqueous solution of hydrogen chloride. It follows that hydrogen chloride is a weaker acid when dissolved in acetic acid than when it is dissolved in water. If it is dissolved in a solvent such as toluene which will not accept protons at all, then there will be no protonated solvent molecules produced, and consequently the solution will not be acidic.

We see, therefore, that whether or not a substance behaves as an acid depends entirely on the nature of the solvent.

When writing equations to represent the dissociation of acids, it is quite common to omit the solvent molecules and to represent the protonated solvent simply by $H^{\oplus}$. This is because the solvent merely acts as a "vehicle" for the proton without affecting its chemistry. Thus the dissociations of hydrogen chloride and sulfuric acid in a solvent are represented simply as

$$HCl = H^{\oplus} + Cl^{\ominus} \qquad \text{and} \qquad H_2SO_4 = 2H^{\oplus} + SO_4^{2\ominus}$$

BASES

These are substances that accept protons. Three common examples are ammonia, the hydroxyl ion, and the oxide ion. They accept protons as indicated in the following equations:

$$\underset{\text{AMMONIA}}{NH_3} + H_3O^{\oplus} = NH_4^{\oplus} + H_2O$$

$$\underset{\text{HYDROXYL ION}}{OH^{\ominus}} + H_3O^{\oplus} = 2H_2O$$

$$\underset{\text{OXIDE ION}}{O^{2\ominus}} + 2H_3O^{\oplus} = 3H_2O$$

The term base is also used for compounds that contain hydroxyl or oxide ions, for example, copper(II) hydroxide, and magnesium oxide, MgO. The alkalis form an important class of bases. These are simply water-soluble bases; common examples are sodium hydroxide, NaOH, potassium hydroxide, KOH, ammonium hydroxide, NH_4OH, and calcium hydroxide, $Ca(OH)_2$. Alkalis turn red litmus blue and neutralize acids; that is, convert hydroxonium ions into water.

$$H_3O^{\oplus} + OH^{\ominus} = 2H_2O$$

or more simply

$$H^{\oplus} + OH^{\ominus} = H_2O$$

The characteristic properties of alkalis are due to the hydroxyl ion $OH^{\ominus}$.

In a neutral solution the concentrations of hydroxonium and hydroxyl ions are equal.

pH meter

The concentration of hydroxonium ions is a measure of the acidity of an aqueous solution. This concentration can be measured directly by using an electrical instrument called a pH meter (Fig. 13.1). The liquid to be measured is placed in the apparatus shown in Fig. 13.1(a), and the pH value is recorded on the meter shown in Fig. 13.1(b). A pH of 4 means that the concentration of hydroxonium ions is 10^{-4} moles per liter, or 1/10,000 moles per liter. A solution with a pH of 1 contains 10^{-1} or 1/10 moles of $H_3O^{\oplus}$ per liter. Thus the lower the pH, the greater the acidity. The degree of acidity or alkalinity of a fluid, that is urine or blood, is often referred to as its *reaction*. A neutral solution contains 10^{-7} moles of $H_3O^{\oplus}$ per liter and therefore has a pH of 7. A solution with an alkaline reaction has a pH greater than 7; one with an acidic reaction has a pH less than 7.

Blood pH

With the introduction of the pH meter, described above, it has become possible to measure the reaction of blood with great accuracy. Using this instrument it has been found that our blood is kept constantly at a pH of 7.35 to 7.42. This is slightly on the alkaline side of neutrality. Only in conditions of sufficient severity to endanger life does the pH fall outside this narrow normal range.

Increase of the pH above 7.42 occurs in alkalosis, for example, from severe vomiting, when there is loss of large quantities of hydrochloric acid, HCl from the stomach, or from excessive alkali therapy which can happen in the treatment of peptic ulcer. Decrease of the pH below 7.35 occurs in acidosis for example in respiratory failure when the lungs fail

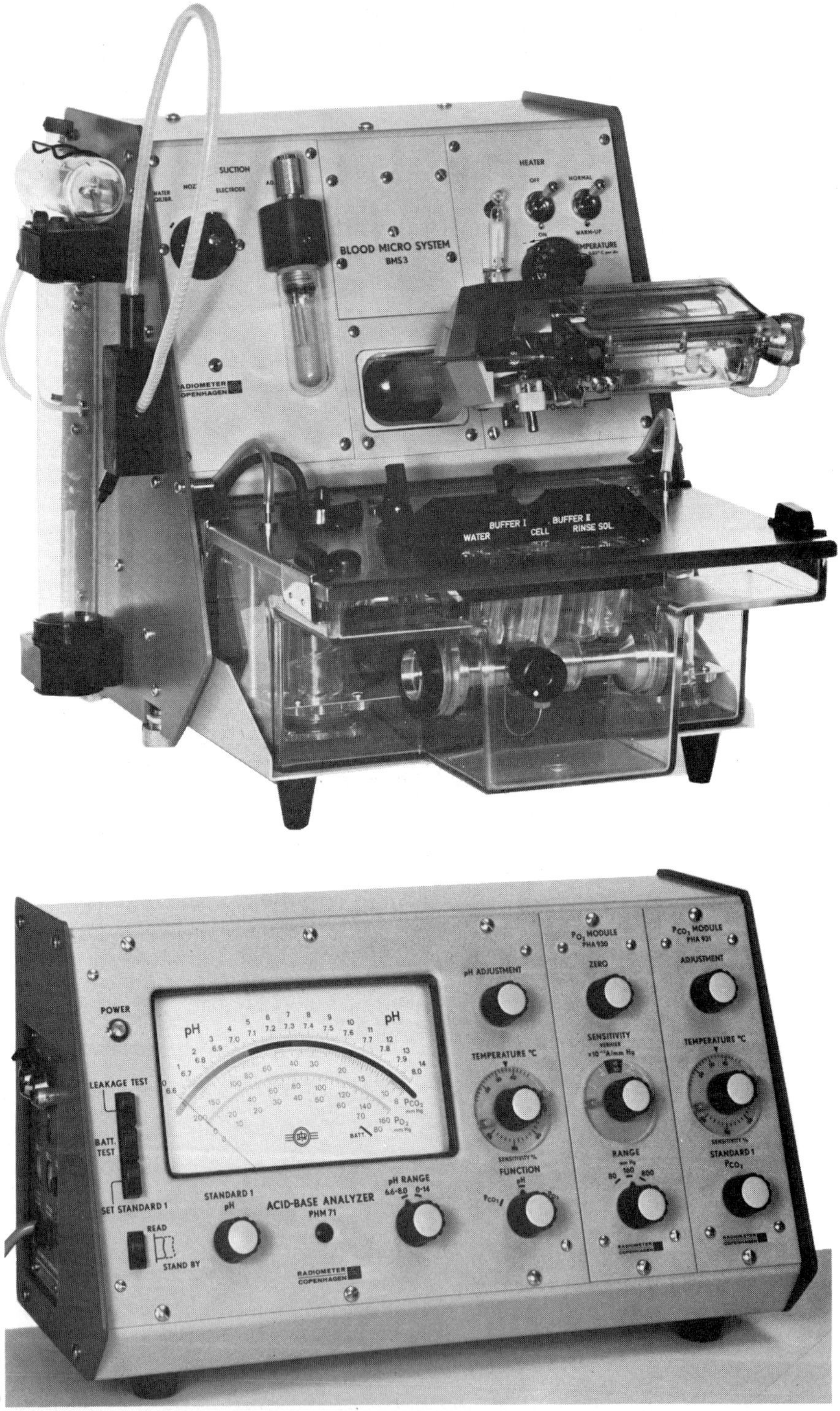

Figure 13.1
Blood pH meter. (Courtesy Radiometer, Copenhagen.)

to expel carbon dioxide, CO_2 rapidly enough to prevent the carbonic acid, H_2CO_3 level in the blood from rising, or in diabetic coma when excessive acids are produced by tissue respiration due to a deficiency of insulin which is essential for the normal combustion of glucose by the tissues.

The constancy of the blood pH in health is maintained by *buffer systems*. When there is a tendency for the pH to fall (that is, an increase in acidity with a rise in hydrogen ion concentration) the respiratory center is stimulated so that more carbon dioxide is expelled, resulting in a lowering of the carbonic acid concentration in the blood and a corresponding rise in pH (that is, an increase in alkalinity). Another buffering system is provided by the kidneys which, provided they are healthy, secrete a more acid urine if there is a tendency to acidosis and a more alkaline urine if there is a tendency to alkalosis. The lungs provide a very quick reaction to change in pH; the kidneys provide a slower control mechanism. In severe kidney diseases too much acid urine may be produced, resulting in renal alkalosis; sometimes too much alkaline urine may be produced, resulting in renal acidosis. Inadequate gaseous exchange by the lungs causes respiratory acidosis, overbreathing causes respiratory alkalosis.

SALTS

Salts are ionic compounds other than acids and bases. They can be made by replacing the hydrogen of an acid by a metal. Thus, magnesium sulfate, $MgSO_4$, Epsom salts, can be prepared by adding magnesium to sulfuric acid.

$$\underset{(s)}{Mg} + \underset{(aq)}{H_2SO_4} = \underset{(aq)}{MgSO_4} + \underset{(g)}{H_2}$$

The hydrogen of the sulfuric acid is replaced by magnesium giving the salt, magnesium sulfate. Hydrogen gas is also formed and bubbles out of the solution causing it to effervesce.

It would be extremely dangerous to prepare sodium sulfate in this way; adding sodium to sulfuric acid causes an explosive reaction. In this case, then, sodium hydroxide or sodium carbonate are added to the acid.

$$\underset{\substack{(aq)\\ \text{SODIUM}\\ \text{HYDROXIDE}}}{2NaOH} + \underset{(aq)}{H_2SO_4} = \underset{(aq)}{Na_2SO_4} + \underset{(l)}{2H_2O}$$

$$\underset{\substack{(s)\\ \text{SODIUM}\\ \text{CARBONATE}}}{Na_2CO_3} + \underset{(aq)}{H_2SO_4} = \underset{(aq)}{Na_2SO_4} + \underset{(l)}{H_2O} + \underset{(g)}{CO_2}$$

Again, the hydrogen of the acid is replaced by a metal, sodium, to form a sulfate. It is possible to make two kinds of salt from it: *sulfates* and *bisulfates*. Sodium sulfate, Na_2SO_4, is prepared by replacing all of the hydrogen of sulfuric acid by sodium. However, if only half the hydrogen is replaced we make sodium bisulfate, $NaHSO_4$.

$$NaOH + H_2SO_4 = NaHSO_4 + H_2O$$

Another acid having two salts is carbonic acid, the weak acid formed when carbon dioxide is dissolved in water.

$$\underset{(l)}{H_2O} + \underset{\substack{(g)\\ \text{CARBON}\\ \text{DIOXIDE}}}{CO_2} = \underset{\substack{(aq)\\ \text{CARBONIC}\\ \text{ACID}}}{H_2CO_3}$$

Carbonic acid has also two replaceable hydrogen atoms per molecule. So it can react with sodium hydroxide to form two different kinds of salt: sodium carbonate, Na_2CO_3 and sodium bicarbonate, $NaHCO_3$. Carbon dioxide, carbonic acid and its salts play an important part in keeping the acid–base balance of our blood constant.

Normal salts and acid salts

The salts formed by replacing all the replaceable hydrogen of an acid are called *normal salts*. Those formed by replacing only part of the replaceable hydrogen are called *acid salts*. We can see then that all acids can form normal salts, but only those with two or more replaceable hydrogen atoms per molecule can form acid salts. The term "acid salt" is only a classification; such a salt does not necessarily give an acidic solution. Thus the acid salt, sodium bicarbonate, $NaHCO_3$ gives a weakly alkaline solution. Examples of normal and acid salts are given in the Table IV.

Note, that of the four hydrogen atoms in the acetic acid molecule, CH_3COOH, only one is replaceable by a metal. Acetic acid, therefore, forms only one kind of salt, a normal salt which is called an *acetate*.

ELECTROLYTES

Acids, bases and salts are all electrolytes. An electrolyte is a substance which when melted or dissolved in water conducts electricity. The salts in our blood are often referred to as the serum (or plasma) electrolytes. Common salt, sodium chloride, is an example of an electrolyte. There is really no such thing as a molecule of sodium chloride. A crystal of salt consists of equal numbers of positively charged sodium ions and negatively charged chlorine ions, packed tightly together (see Fig. 5.3).

Table IV. Normal and Acid Salts

ACID	FORMULA	NUMBER OF REPLACEABLE HYDROGEN ATOMS PER MOLECULE	FORMULA OF SODIUM SALTS	NAME	TYPE OF SALT
hydrochloric	HCl	1	NaCl	sodium chloride	normal
nitric	HNO_3	1	$NaNO_3$	sodium nitrate	normal
sulfuric	H_2SO_4	2	Na_2SO_4	sodium sulfate	normal
			$NaHSO_4$	sodium bisulfate	acid
carbonic	H_2CO_3	2	Na_2CO_3	sodium carbonate	normal
			$NaHCO_3$	sodium bicarbonate	acid
phosphoric	H_3PO_4	3	Na_3PO_4	sodium phosphate	normal
			Na_2HPO_4	sodium monohydrogen phosphate	acid
			NaH_2PO_4	sodium dihydrogen phosphate	acid
acetic	CH_3COOH	1	CH_3COONa	sodium acetate	normal

When salt is dissolved in water the sodium ions and the chlorine ions move about almost independently of each other. The ions of the electrolytes in the blood move about similarly and the concentration of each type of ion is maintained at a separate level.

Blood electrolytes

The main electrolyte ions in the blood are as follows:

Cations or positive ions	sodium	$Na^{\oplus}$	133–144 mmol/liter (m.eq./liter)
	potassium	$K^{\oplus}$	3.5–5.5 mmol/liter (m.eq./liter)
Anions or negative ions	chlorine	$Cl^{\ominus}$	96–106 mmol/liter (m.eq./liter)
	bicarbonate	$HCO_3^{\ominus}$	23–31 mmol/liter (m.eq./liter) [adults] 18–23 mmol/liter (m.eq./liter) [children]

The concentrations are expressed in millimoles per liter. This is now the recommended method of expressing concentration so as to represent clearly the relative proportion of each of the different ions. For the electrolytes, the levels are numerically the same as for the obsolescent term milli-equivalents per liter.

Estimation of the electrolytes is of great importance in cases of dehydration, for example, from diarrhea, vomiting, burns or excessive sweating, so that when the lost fluid is replaced, each of the various ions may be maintained at the correct level. Their estimation is equally important in cases of fluid retention which causes waterlogging of the tissues, known as edema; this is seen in heart failure, kidney failure, liver failure, and starvation.

The electrolyte balance is also upset when the function of certain endocrine glands is disturbed, for example, diabetic ketosis (failure of the islets of Langerhan's in the pancreas to produce enough insulin) or Addison's disease (failure of the adrenal cortex to produce enough steroids).

Steroids are now used in the treatment of such conditions as asthma, rheumatoid arthritis, eye inflammations and certain skin diseases. Steroid treatment can cause disturbance of the plasma electrolytes as a side-effect which must therefore be checked regularly during such therapy.

Tissue electrolytes

The concentration of electrolyte ions present in the blood plasma or serum is practically identical with that in tissue fluid; but the concentration of the various types of ions in the intracellular fluid is very different.

The main difference between the intracellular and extracellular fluid lies in the concentration of sodium and potassium ions. In the plasma, there is a much higher concentration of sodium than potassium ions. Inside the cells this ratio is reversed.

The high concentration of potassium inside cells has one important practical implication. If special care is not taken in collecting blood for electrolyte examination the red cells may be damaged and the potassium from within the cells be released into the plasma, giving an erroneously high reading for the potassium concentration.

To avoid damaging the red cells (hemolysis), the syringe and needle must be dry, as in the sterile disposable packs; the blood must not be squirted out through the needle, and the collecting container must not contain water but may, of course, contain liquid anticoagulant. In addition the estimations must be undertaken soon after the blood is collected. If blood is allowed to stand for several hours the potassium ions gradually diffuse out of the red cells into the plasma, again giving a false high reading for the potassium concentration, even without any hemolysis.

14

Osmosis and Filtration

OSMOSIS

When red blood cells are put into plain water they immediately burst. This is because they are overdistended by water passing into them through their cell membrane. This passage of water is called osmosis and the pressure which bursts the red blood cells is called osmotic pressure (Fig. 14.1(a)). However, if our red blood cells are introduced into a 2% saline solution, a salt concentration significantly higher than that existing inside the cells, the opposite happens. Water passes from the red blood cells into the more concentrated salt solution outside, causing them to shrink and crenate (Fig. 14.1(b)).

These are two examples of the effect of osmosis. For osmosis to occur there must be a membrane which allows water to pass through but which holds back the molecules or ions of substances dissolved in the water. This is called a semipermeable membrane. In the illustrations the red cell envelope is the semipermeable membrane. Solvent (water in the case of cells) is always transferred from the weaker to the more concentrated solution, unless of course we apply sufficient pressure to coun-

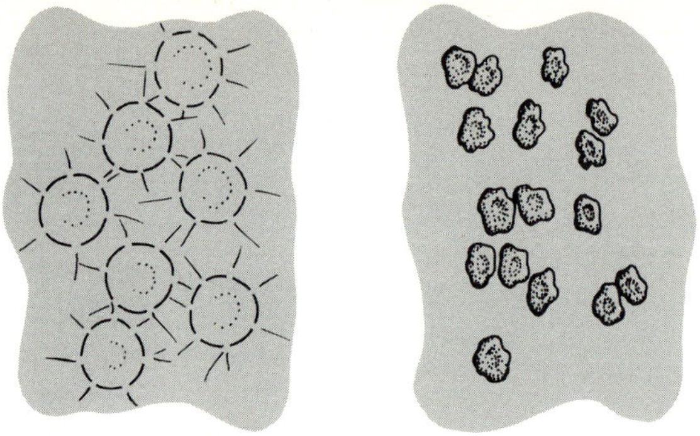

(a) (b)

Figure 14.1
(a) Cells bursting in water; (b) cells crenating in two percent saline.

teract the osmotic pressure. The exact mechanism of osmosis is not yet fully understood but this does not prevent us from understanding its practical implications.

To demonstrate osmosis on a larger scale a piece of parchment may be used as a semipermeable membrane. This can be fixed over the open end of a thistle funnel after filling the funnel with sugar solution (Fig. 14.2). If the funnel is then inverted with the parchment under the water and the tap opened, the level of water will rise in the stem of the thistle funnel. The height to which it rises is a measure of the osmotic pressure. The greater the concentration of particles (molecules or ions) dissolved, the greater the osmotic pressure produced.

Osmotic pressure in the body

An osmotic pressure only develops if the semipermeable membrane will not allow the dissolved salts to pass through. The walls of our blood capillaries only partially prevent the salts (electrolytes) in the blood from diffusing out into the tissue spaces. Increasing the concentration of electrolytes in the blood only temporarily produces an osmotic effect, drawing fluid into the capillaries from the tissue spaces. Even this temporary effect can be of value, for example, in reducing intracranial pressure to facilitate operations on a brain tumor. Within a few hours however there is sufficient diffusion of electrolytes across the capillary walls for the salt concentrations to equalize.

The plasma proteins have large molecules and, except in inflammation, they cannot traverse the capillary walls. So it is the plasma proteins which have the most osmotic effect in relation to the tissue fluid. The two main types of protein in the plasma are albumin and globulin. Globulin molecules are bigger. But there are far more molecules of al-

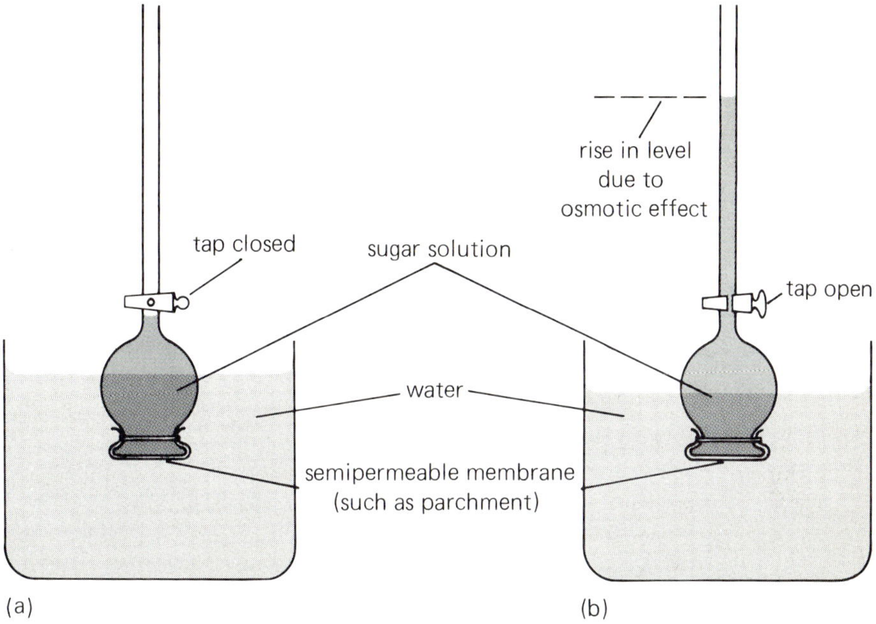

Figure 14.2
(a) Apparatus to show osmotic effect of semipermeable membrane; (b) when the tap is open the water molecules pass through the semipermeable membrane but the sugar molecules do not.

bumin. Since the osmotic pressure is proportional to the number of molecules it is the albumin which produces the greatest osmotic effect.

The blood in our arteries is under considerable pressure. At the arterial end of our capillaries the pressure is still sufficient not only to prevent osmotic flow into the capillary but actually to cause fluid to pass out through the capillary wall into the tissue spaces (Fig. 14.3). By the time blood reaches the venous end of the capillary its pressure has dropped so much that the osmotic effect of the plasma proteins inside the capillary is no longer counteracted and fluid returns into the vessel again, the capillary wall acting as a semipermeable membrane.

If the plasma protein (especially the albumin) is reduced, as in nephritis or protein starvation, the osmotic pressure is correspondingly reduced until eventually it is insufficient for fluid to pass back into the capillary from the tissue spaces. So fluid accumulates in the tissues causing edema (Gr. *oidema*, swelling), especially in the dependent parts and also in lax tissues around the eyes.

In congestive cardiac failure the heart fails to pump the blood from the veins to the arteries fast enough. So there is a pressure build-up in the veins. As a result of this venous engorgement the pressure at the venous end of the capillary loops is higher than normal. The osmotic

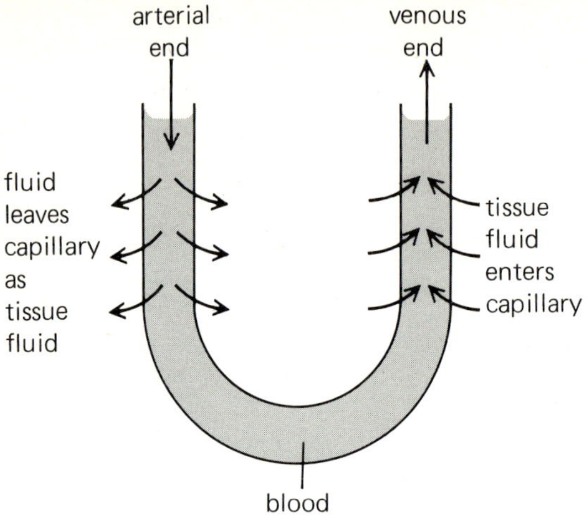

Figure 14.3
Capillary loop.

pressure due to the plasma proteins in the capillary is insufficient to overcome this raised capillary pressure. So, again, fluid accumulates in the tissue spaces and edema develops, particularly where the effect of gravity is most marked, as in the ankles, or if the patient is sitting in bed, over the sacrum.

Tonicity

If two solutions have the same osmotic pressure they are called isotonic. In medicine the term "isotonic saline" is used to describe a salt solution (0.85% sodium chloride in water) which has the same osmotic pressure as the salts in the blood. Ringer-Locke's solution is similarly "isotonic" but has a multiple salt composition more closely resembling that actually present in the blood. Red blood cells neither shrink nor burst when placed in isotonic saline.

Recent developments have made it possible to measure osmotic pressure directly. The apparatus used is called an osmometer and it works by measuring the freezing point depression produced by a substance in water. The results are expressed in milli-osmoles/liter. Thus 1000 milli-osmoles of a substance dissolved in 1 liter of water depresses the freezing point by 1.86°C. The normal osmolarity of blood serum is 275–295 milli-osmoles/liter (mOsm/liter) and is mainly due to the electrolytes.

The term "hypertonic saline" is used to describe a solution having a salt concentration which is greater than that in blood, for example, 10% saline (3200 mOsm/liter). The raised intracranial pressure of cerebral tumor may be temporarily lowered by injecting hypertonic solutions, e.g. 10% saline or 25% glucose (1400 mOsm/liter) intravenously. The osmotic effect is sufficient to draw fluid out of the tissue spaces for 2 to 4 hours, causing the brain tissue to shrink markedly and allow the cranial bones to be removed without damage to the brain which would otherwise bulge out due to the raised incranial pressure.

The term "hypotonic saline" is used to describe a solution having a salt concentration weaker than that in blood. It is used in conditions where all the constituents of the blood have become over-concentrated (hemo-concentration) due to water loss without corresponding salt loss.

MEDICAL APPLICATIONS OF OSMOTIC PRESSURE

Saline purgatives

The best known saline purgative is magnesium sulfate ($MgSO_4$ or Epsom salts). To produce the maximum effect it should be taken on an empty stomach, usually first thing in the morning, so that it passes directly through the stomach into the duodenum. It should be dissolved in sufficient water to be isotonic or slightly hypotonic. A 6.5% solution is isotonic. So 6.5 g will retain 100 ml of water in the gut, which should nearly double the volume of the feces. It is important not to give magnesium sulfate in hypertonic solution as this induces closure of the pylorus and may cause vomiting.

Osmotic diuretics

Any increase in the amount of substances dissolved in the urine causes an increase in the osmotic pressure of the urine and results in an increase in the amount of urine produced. Two substances which act in this way and which can be taken by the mouth are urea and potassium nitrate. They should not, however, be used if there is kidney damage.

Another substance acting in a similar way is sodium sulfate, given in 6% solution as an intravenous drip. It is used when the kidneys have stopped functioning following an operation on the urinary tract.

Osmotic poultice

Before the introduction of antibiotics the standard treatment of boils and carbuncles was the application of a poultice consisting of magnesium sulfate and glycerine. The effect of this is to draw out fluid, including pus, from the affected site. This treatment can still be used with advantage where the lesion fails to resolve on antibiotic therapy.

FILTRATION

A simple example of filtration is a tea strainer. This has a mesh with holes small enough to hold back the tea leaves but which readily allows the tea infusion to pass through. A Tetley's tea bag acts in the same way.

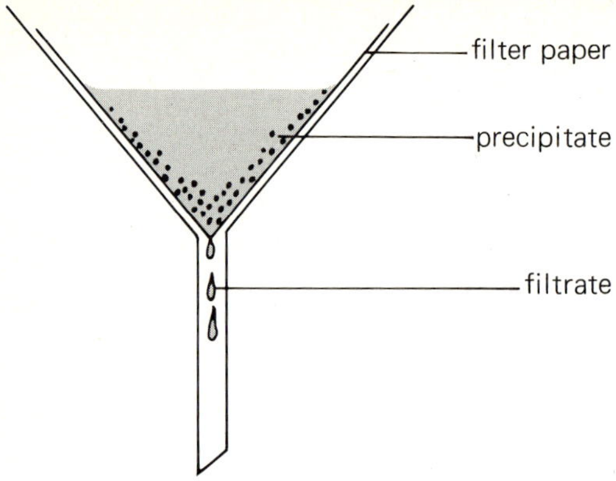

Figure 14.4
Filtration funnel.

So does the chemist's filter paper placed in a filtration funnel (Fig. 14.4) which holds back a precipitate but allows fluid and dissolved substances to pass through.

One medical application of this principle is the bacteriological filter (Fig. 14.5). This has pores which are too small for bacteria to pass through. This enables solutions that would be damaged by heat to be rendered bacteria-free without autoclaving. It should be noted that viruses, being signifiantly smaller than bacteria, can pass through such filters.

The pores in a bacteriological filter are so small that considerable pressure is required to force the fluid through. This is provided by

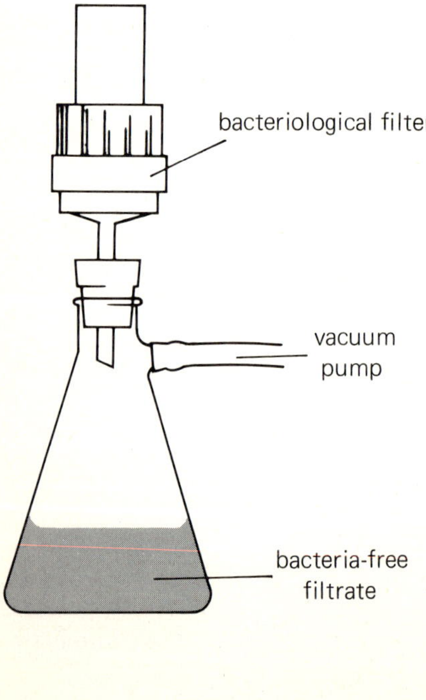

Figure 14.5
Sterilization by bacteriological filtration.

means of a vacuum pump. By evacuating the flask into which the fluid is being filtered (Fig. 14.5) the atmospheric pressure effectively forces the fluid through.

Filtration in the kidney

In the kidney, the filtration force is provided by the blood pressure (Fig. 14.6). The filtration occurs in the renal glomeruli of which there are about a million in each kidney. The filter in each glomerulus is essentially composed of two layers, the capillary wall and the lining of Bowman's capsule (Fig. 14.7).

The pressure within the glomerular capillaries is sufficient to force through this double-layered filter the fluid component of the blood including all the substances dissolved in it except those with large molecules like the plasma proteins. Only if the kidney is damaged, as in nephritis, do the proteins pass through; albumin is the smallest protein

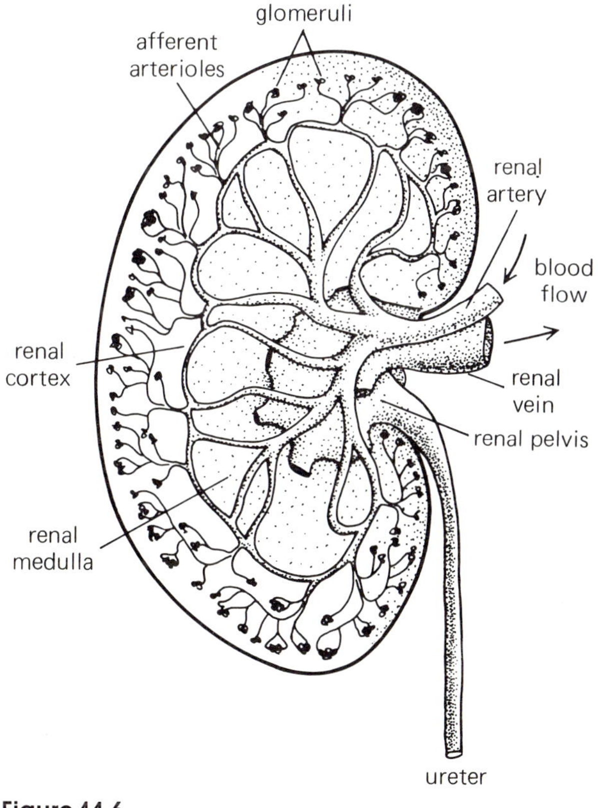

Figure 14.6
Kidney showing how branches of the renal artery supply blood under pressure to the glomeruli in the renal cortex through the afferent arterioles.

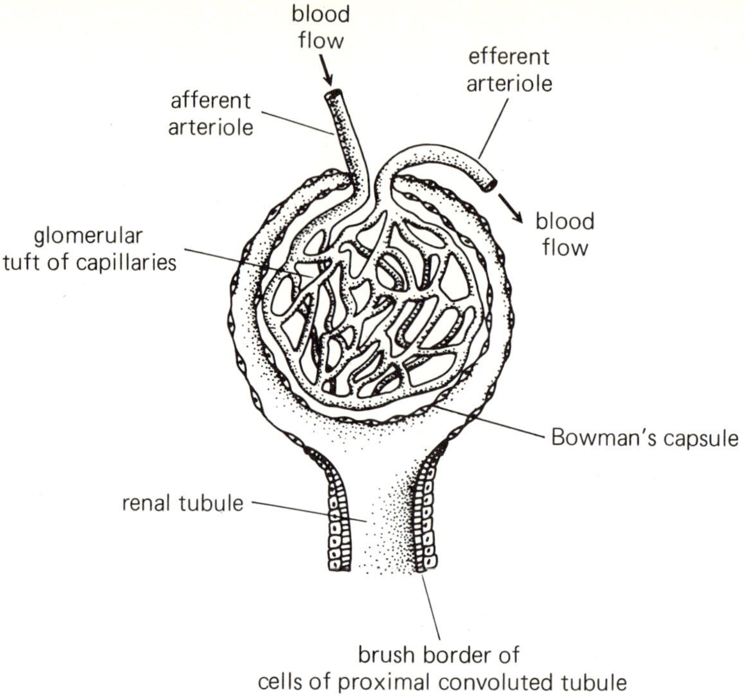

Figure 14.7
Glomerulus and Bowman's capsule. (L. *afferens*, carrying toward and L. *efferens*, carrying from.)

and evidence of such damage is the presence of albuminuria (albumin in the urine).

This filtration process is the first step in the formation of urine. Of the 200 liters of water that are passed through the glomeruli every day, all but 1.5 liters are reabsorbed by the kidney tubules (Fig. 14.8). Nearly all the sodium and chloride is reabsorbed and normally all the glucose is reabsorbed. This selective reabsorption of those chemicals required by the body is an active process carried out by the renal tubule cells and involves work against osmotic pressure. Water reabsorption is controlled by the antidiuretic hormone of the posterior pituitary gland; the more hormone that is secreted, the more water is reabsorbed.

The waste materials filtered through the blood by the kidney are described in the section on waste materials in Chapter 29 and the one on urine in Chapter 30.

Artificial Kidney

Anuria (failure of the kidney to produce urine) occurs following shock and its associated fall in blood pressure. It may result from an incompatible blood transfusion. The artificial kidney provides a means of tiding the patient over the crisis until kidney function recovers. It is also used

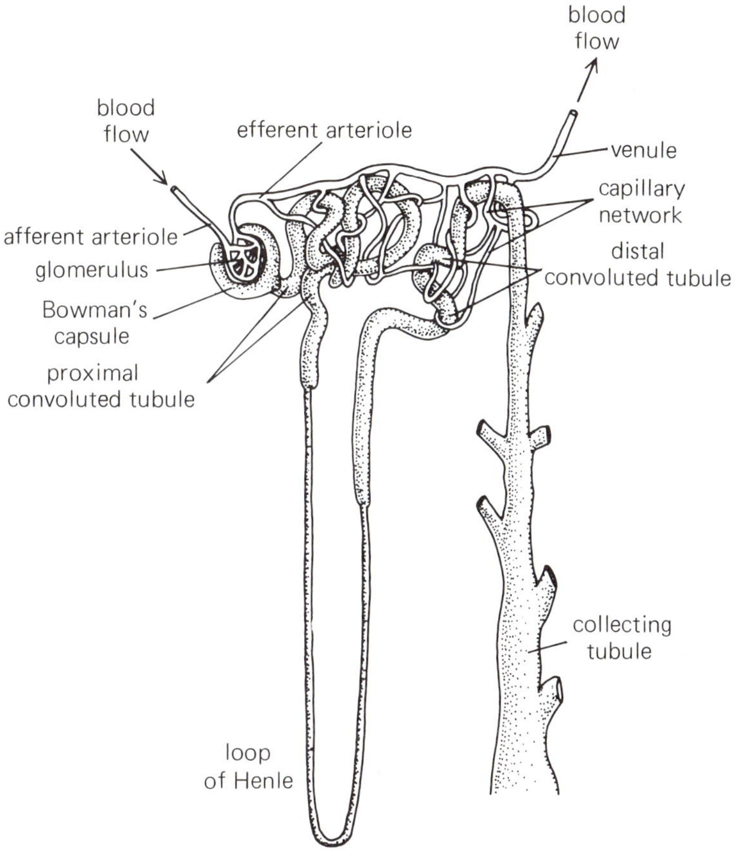

Figure 14.8
Kidney tubule: each renal unit (nephron) consists of a glomerulus, a Bowman's capsule, and a tubule. Each kidney tubule consists of a proximal convoluted tubule, a loop of Henle, and a distal convoluted tubule, which then opens into a collecting tubule, each of which drains a number of kidney tubules into the renal pelvis.

in more permanent cases of kidney failure, particularly prior to renal transplant. In the artificial kidney the patient's blood is passed through long coils of collodion tubing surrounded by a saline solution. These tubes are sufficiently permeable for all but the larger molecules, such as plasma proteins, to pass through. A similar effect is produced by "peritoneal dialysis" in which a saline solution is passed constantly through the peritoneal cavity. It is important to monitor the serum electrolytes in these patients.

15

Sodium and Potassium

Sodium and potassium are the most common elements in the chemical group known as the alkali metals. They are also the two most abundant metals in the body and the ones most frequently estimated in blood samples. They are present in the form of sodium ions $Na^{\oplus}$ and potassium ions $K^{\oplus}$.

ATOMIC STRUCTURE

All the alkali metals have a similar atomic structure in having only one electron in the outer shell, which gives each element a valency of one. Diagrammatically the first three elements in the series may be represented thus:

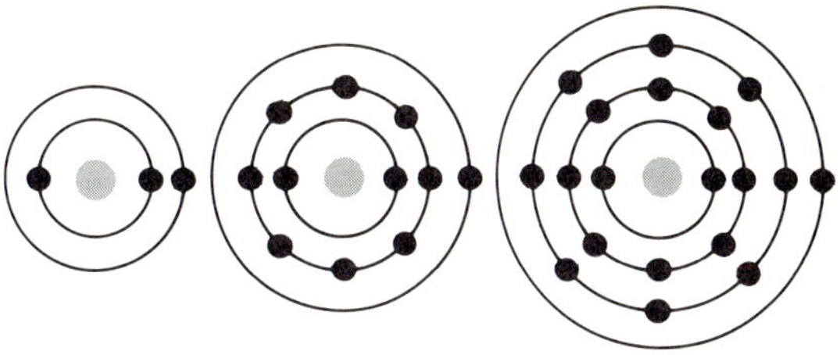

	LITHIUM	SODIUM	POTASSIUM
Electron configuration	2, 1	2, 8, 1	2, 8, 8, 1
Atomic number	3	11	19

The above three elements are all of medical importance, lithium in the treatment of certain mental disorders, sodium and potassium as key elements in the normal body chemistry.

If the sodium atom loses its outer (valence) electron it becomes a positively charged sodium *ion* $Na^{\oplus}$. Since the two remaining shells are completely filled it now has the same electron configuration as the inert gas neon and is very stable. In contrast the sodium *atom* is very unstable, that is, very reactive. The effect of the complete inner electron shells is to screen the solitary electron in the outer shell from the electrical attraction of the nucleus of the atom. As this outer electron is held in orbit by a comparatively weak force it can be transferred to the atom of another element, such as chlorine, and an ionic link is established between the ions produced.

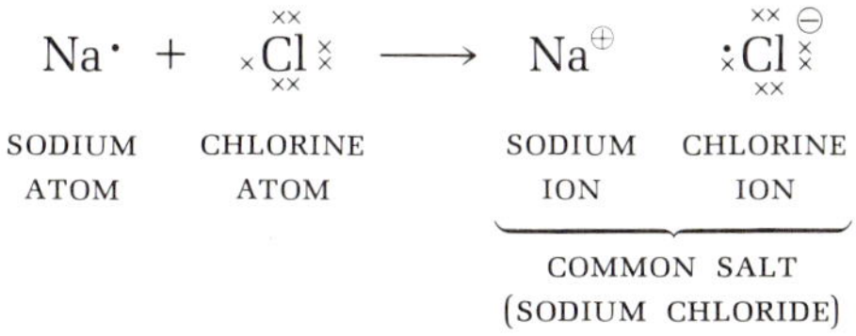

The compounds of sodium are not composed of molecules but of positive and negative ions, attracted toward each other by their opposite charges. The electrolytes in our body fluids are of this nature.

With potassium the structural factors promoting reactivity are even more pronounced. The lone outer (valence) electron of potassium is further away from the nucleus than that of sodium; it is also more effectively screened from the electrical attraction of the nucleus of the atom because of the additional inner ring of eight electrons. So the outer electron is even more readily transferred to another element, making potassium still more reactive than sodium.

Lithium on the other hand is generally less reactive than sodium, as might be inferred from its atomic structure. The outer solitary electron is closer to the atomic nucleus and much less effectively screened from the nucleus by the intervening duplet of electrons. It is therefore more firmly bound to the lithium atom and less readily transferred to the atom of another element.

The readiness with which sodium and potassium form compounds and the difficulty with which the compounds are separated into their components is the main reason why the discovery of these elements was relatively late in chemical history.

DISCOVERY

Potassium was the first of the alkali metals to be discovered, by Humphry Davy, in England in 1807. Davy is better known for his discovery of the anesthetic "laughing gas" (nitrous oxide, N_2O) and for his invention of the miner's safety lamp.

You will remember that if an electric current is passed through an electrolyte, it decomposes, the process being called electrolysis. Davy used electrolysis to decompose caustic potash, KOH. At the cathode, the negative pole, he saw "small globules of a high metallic luster." He showed that these globules were in fact a metal, which he called potassium, and that it had properties which were very unusual for a metal, being lighter than water and reacting with it in a spectacular manner.

In the same year (1807) Davy discovered sodium, by electrolysis of caustic soda, NaOH.

OCCURRENCE

Sodium is the sixth and potassium the seventh most abundant element on earth, each constituting about 2.5% of the earth's crust. Sodium is the commonest of the alkali metals and potassium the next most common.

Both elements occur in igneous rocks, such as felspar (sodium aluminum silicate) and potash felspar (potassium aluminum silicate). Weathering gradually breaks down the rocks. The insoluble felspars are partly converted by the carbon dioxide in the air to soluble carbonates. Eventually far more sodium than potassium is converted into soluble carbonate. This is washed away and accumulates in the oceans or inland seas. As a result there is far more sodium than potassium in the sea.

Of the dissolved solids in ordinary sea water, 75% is sodium chloride and less than 3% is potassium chloride. Even so, some of the richest deposits of potassium salts are at the sites of ancient lakes or sea beds, Stassfürt in Northern Germany being one of the largest. Deposits of sodium chloride or rock salt are found in almost every country; one of the most renowned is in Siberia.

In plants potassium occurs as salts of organic acids, for example oxalates in sorrel and rhubarb. Wood ash contains about 30% potassium carbonate, a fact much appreciated by gardeners. Extraction with water and evaporation in metal pots gave rise to the name potash (pot ash).

The body fluids of all animals contain sodium and potassium ions. Sodium is present in very much the higher concentration in the extracellular fluids including blood. Potassium predominates in the intracellular fluid. The milk of carnivorous animals contains sodium and potassium in equal amounts; the milk of herbivorous animals and of humans contains a higher proportion of potassium, their milk ash containing about 5% sodium and just over 20% potassium.

PROPERTIES

Sodium and potassium are soft, silvery white metals, sufficiently light to float on water; lithium is the lightest of all metals but has a harder consistency. At room temperature sodium and potassium are soft enough to be cut with a knife, squeezed between the fingers (not recommended) or extruded through an orifice, potassium being the softest of the three. When heated out of contact with air they melt into silvery liquids.

The chemical reactions of the three metals are similar, those of potassium being more vigorous than those of sodium which in turn are more vigorous than those of lithium. They are not attacked by perfectly dry oxygen but in moist air each rapidly tarnishes to form the metal hydroxide and carbonate. When heated in air or oxygen the metals burn to form oxides.

Reaction with water

Lithium, sodium, and potassium react with water to form the hydroxide and hydrogen gas.

$$\underset{\text{SODIUM}}{\underset{(s)}{2Na}} + \underset{(l)}{2H_2O} = \underset{\text{SODIUM HYDROXIDE}}{\underset{(aq)}{2NaOH}} + \underset{\text{HYDROGEN}}{\underset{(g)}{H_2}}$$

A small portion of sodium, placed on water in an open vessel, shoots about the surface. The heat produced is not sufficient to ignite the hydrogen spontaneously unless the sodium is prevented from moving. If it is ignited, it burns with the yellow flame characteristic of sodium. Lithium reacts with water similarly but less vigorously. If the hydrogen is ignited it burns with a crimson lithium flame.

A pellet of potassium placed on water reacts even more vigorously than sodium and the hydrogen ignites spontaneously.

$$\underset{(s)}{2K} + \underset{(l)}{2H_2O} = \underset{(aq)}{2KOH} + \underset{(g)}{H_2}$$

The hydrogen produced burns with the lilac color characteristic of potassium. In view of the readiness with which sodium and potassium react with water, they need to be stored under oil. Even the small amount of water vapor in the air is sufficient to start a reaction.

Reaction with acids

The metals react strongly and, with the exception of lithium, explosively with acids to form salts and release hydrogen.

$$\underset{(s)}{2Na} + \underset{(aq)}{2HCl} = \underset{(aq)}{2NaCl} + \underset{(g)}{H_2}$$

Reaction with other elements

Potassium combines even more readily than sodium with chlorine, oxygen, and sulfur; lithium combines rather less readily. Sodium and potassium have an affinity for these elements which is so great that they will decompose compounds of the above elements to combine with them, for example,

$$\underset{\text{ALUMINUM CHLORIDE}}{AlCl_3} + \underset{\text{SODIUM}}{3Na} = \underset{\text{SODIUM CHLORIDE}}{3NaCl} + \underset{\text{METALLIC ALUMINUM}}{Al}$$

Sodium and potassium are thus powerful reducing agents; each atom readily donating an electron and becoming converted into sodium and potassium ions in the process.

SODIUM AND POTASSIUM IONS

Sodium chloride, NaCl

Sodium chloride occurs either as colorless crystals or in the form of a white crystalline powder as in table salt. It is odorless but has a characteristic salty taste. A solution of sodium chloride in water is referred to simply as saline.

Normal sodium chloride solution, in the sense used by chemists, contains the gram formula weight (58.5 g) in one liter of water. But in a medical context the term "normal saline" refers to physiologically normal saline. This denotes a saline solution which is isotonic with human and animal tissues.

Physiologically normal saline is 0.9% NaCl and its freezing point is 0.55°C below that of water. You will recall from the last chapter that a freezing point depression of 1.86°C is produced by a solution with an osmolarity of 1000 mOsm/liter. Experiment shows that osmolarity is proportional to freezing point depression. Therefore, 0.9% NaCl has an osmolarity of $1000 \times \frac{0.55}{1.86} = 296$ mOsm/liter.

Physiologically normal saline is extensively used for the cleansing of wounds. Being isotonic with living tissue it does not disturb the osmotic equilibrium of the cells and does not cause the stinging sensation that water may produce on cuts and bruises. Intravenous preparations of sterile normal saline are frequently used for transfusion (Fig. 15.1), as in the treatment of dehydration due to severe gastroenteritis or cholera. Its U.S. pharmaceopeial (USP) category is "electrolyte replenisher."

Hypertonic saline solutions are used in dressing packs for the treatment of varicose ulcers. The effect of hypertonic saline is to remove fluid from the tissues osmotically, thus reducing edema (excess fluid in the tissues). An example of hypertonic saline commercially available is 1.8% NaCl which has a freezing point depression of 1.10°C and therefore an osmolarity of $1000 \times \frac{1.10}{1.86} = 591$ mOsm/liter, nearly twice that of living tissue.

As the saline concentration increases, the osmolarity becomes larger. Thus, 3% NaCl has a freezing point depression of 1.82°C and hence an osmolarity of $1000 \times \frac{1.82}{1.86} = 978$ mOsm/liter. This is over three times greater than the osmolarity of living tissue. So saline of this

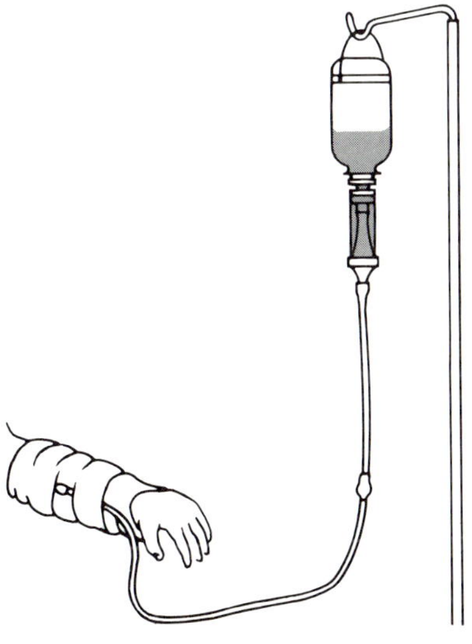

Figure 15.1
Saline transfusion apparatus.

strength has a powerful osmotic effect in removing excess fluid from tissues. As fluid is withdrawn from the tissues it tends to dilute the hypertonic saline, gradually reducing its osmolarity and making it less effective. So a hypertonic saline pack is most effective just after it is applied.

Tablets of sodium chloride which have been "enteric coated" to prevent digestion in the stomach are given by mouth in the treatment of salt loss, for example for heat exhaustion in the tropics. Strong salt solution by mouth is an emetic (causes vomiting). It acts by irritating the stomach. Like other emetics it should not be used in the treatment of poisoning by substances such as acids, which can damage and weaken the stomach wall, because of the danger of perforation. Sodium chloride can be highly toxic or even fatal in large doses as when given to infants mistakenly as sugar.

Potassium chloride, KCl

Potassium chloride is a white crystalline powder resembling common salt and having a similar saline taste. Its main use in medicine is in the treatment of potassium depletion. Its USP category is "electrolyte replenisher." It may be given by mouth as a mixture or as a solution produced from effervescent potassium chloride tablets. Enteric-coated tablets of potassium chloride are also used but need to be of the slow-release type to avoid ulceration of the bowel wall by the high local concentration of potassium chloride. Potassium chloride is sometimes given intravenously, well diluted in physiologically normal saline or 5% dextrose. If injected too rapidly it can depress the heart's action or even cause cardiac arrest by "potassium inhibition."

Sodium iodide, NaI

Sodium iodide is a white crystalline powder with a slightly bitter saline taste. In moist air it decomposes, turning yellow due to the liberation of iodine. It is used for preventing goiter in regions where this is endemic, being given as 0.01% sodium iodide in table salt. Its USP category is "source of iodine." In hospital practice it is used as premedication before a thyroid scan to study thyroid activity, also prior to operation for removal of thyroid gland tissue in cases of thyrotoxicosis.

Potassium iodide, KI

Potassium iodide occurs as slightly opaque crystals or as a white crystalline powder. It is odorless with a slightly bitter saline taste, similar to sodium iodide. It is the most commonly used form in which iodine is administered in regions where goiter is endemic due to iodine lack. It is taken as 0.01% potassium iodide in table salt.

In medical practice potassium iodide is also used as an expectorant, to thin down the bronchial mucus in cases of asthma and bronchitis. Its USP category is "expectorant."

Sodium bicarbonate, $NaHCO_3$

Sodium bicarbonate, also called bicarbonate of soda, occurs either as small opaque crystals or as a white powder. It is odorless but has a salty taste. It is one of the more commonly used gastric antacids but has the disadvantage that it liberates carbon dioxide in the stomach.

$$\underset{(s)}{NaHCO_3} + \underset{(aq)}{HCl} = \underset{(aq)}{NaCl} + \underset{(l)}{H_2O} + \underset{(g)}{CO_2}$$

Although giving rapid subjective relief by neutralizing acidity, often with satisfyingly noisy belching due to the carbon dioxide gas released, it tends to stimulate the stomach to produce more acid. On the other hand if too much sodium bicarbonate is taken it can cause alkalosis

Used both orally and intravenously, sodium bicarbonate solution is used in the treatment of acidosis as for diabetic coma. Its USP category is "antacid."

Lithium carbonate, Li_2CO_3

Lithium carbonate is a white amorphous or microcrystalline powder. It is odorless but has a slightly alkaline taste. Medically it is used in the treatment of mental patients and recently there has been a great increase in the use of lithium carbonate for the treatment of affective disorders (mania and depression) with encouraging results. In order to maintain the blood concentration of lithium at the correct level (0.6 to 1.5 mmol/liter) it is convenient for the treatment to be taken orally in the form of capsules or tablets, for example, Eskalith (Smith, Kline and French) and Lithonate (Rowell Labs., Inc.), which provide a gradual release of lithium carbonate into the small intestine.

Sodium hydroxide, NaOH

A powerful and poisonous alkali, sodium hydroxide, also known as caustic soda, is a white, slightly translucent solid. On exposure to air it first becomes moist; then the surface solution reacts with carbon dioxide in the air to form a dry crust of sodium carbonate.

Sodium hydroxide is an ionic compound. It readily dissolves in water producing a solution of sodium and hydroxyl ($OH^{\ominus}$) ions. The high concentration of hydroxyl ions makes the solution strongly alkaline (and highly poisonous), enabling it to neutralize acids.

$$\underset{\text{BASE}}{\underset{(aq)}{NaOH}} + \underset{\text{ACID}}{\underset{(aq)}{HCl}} = \underset{\text{SALT}}{\underset{(aq)}{NaCl}} + \underset{\text{WATER}}{\underset{(l)}{H_2O}}$$

If the equation is written ionically, thus

$$Na^{\oplus} + OH^{\ominus} + H^{\oplus} + Cl^{\ominus} \longrightarrow Na^{\oplus} + Cl^{\ominus} + H_2O$$

we can see that the sodium and chlorine ions take no part in the reaction which is simply the combination of hydrogen and hydroxyl ions to form water, namely $H^{\oplus} + OH^{\ominus} \rightarrow H_2O$. The USP category of sodium hydroxide is a "pharmaceutic aid (alkalizing agent)."

Potassium hydroxide, KOH

Potassium hydroxide, sometimes known as caustic potash, has properties very similar to sodium hydroxide and, like it, is a powerful and poisonous alkali. Its USP category is also as a "pharmaceutic aid (alkalizing agent)." Concentrated caustic potash solution is used in biology for dissolving dead flesh away from the bone, for instance to obtain the skeleton of a frog.

Sodium and potassium in the body

In the body the sodium ion, $Na^{\oplus}$ is the most abundant cation in the extracellular fluid and the potassium ion, $K^{\oplus}$ predominates in the intracellular fluid. Of the total potassium in the body 98% is situated intracellularly, the concentration being highest in the cells of the brain. Intracellular fluid is maintained at a very constant composition. This constancy is essential for life. During starvation, for example, it is not until the very late stages that the body begins to break down intracellular materials for fuel with a resultant increase in potassium excretion in the urine.

The concentration of sodium and potassium in the extracellular fluids is also maintained within very narrow limits, although not quite as constant as in the intracellular fluid. Thus the normal serum sodium is within the range 133 to 144 mmols/liter and the potassium 3.5 to 5.5 mmols/liter. To maintain these cations at the right level requires a flexible control mechanism to balance the losses in urine, perspiration, and feces against the dietary intake.

The dietary intake of sodium varies very considerably, usually being between 1 to 6 g (approximately 50 to 300 mmol) daily. Major factors are the amounts of cooking and table salt added to food. As a buffer against the variations in dietary intake there is a substantial reservoir of sodium provided by the bony skeleton. Approximately one-third of the body sodium is in skeletal bone and can be rapidly mobilized when required.

The chief regulator in maintaining a constant extracellular sodium concentration is the kidney. The excretion of sodium in the urine is adjusted very closely to the body's needs. This is achieved by the extent to which sodium is reabsorbed in the kidney tubules. This reabsorption is controlled by steroid hormones from the adrenal cortex, especially aldosterone.

If sufficient adrenal cortex tissue is damaged, as in Addison's disease, the steroid secretion becomes insufficient to promote the necessary reabsorption of sodium by the kidney tubules. Excessive sodium is lost, resulting in a low sodium level in the blood and tissue fluid. This is called hyponatremia (Gr. *hypo,* low, +Gr. *natrium,* sodium, +Gr. *haima,* blood). When the serum sodium falls to about 120 mmol/liter mental confusion occurs. Below 110 mmol/liter lethargy, coma and convulsions develop. Without prompt treatment as with intravenous hypertonic sodium chloride, death may result.

The reverse occurs in tumors or overactivity of the adrenal cortex, as in Cushing's syndrome. Adrenal steroids are secreted in excess, causing the kidneys to reabsorb too much sodium. This results in water retention as an osmotic effect and the tissues become waterlogged or edematous. Steroids used in therapy, for example, for rheumatoid arthritis, can have a similar effect. Failure of the kidneys to excrete sufficient sodium ions is an important factor in the edema of heart and kidney failure.

The dietary intake of potassium is about 4 g or 100 mmol/day. Part of this (5–10%) is lost in the feces. The remainder is excreted by the kidneys. Reabsorption of potassium by the kidney tubules is never as complete as that of sodium, even in states of severe potassium depletion.

Potassium depletion may be due to diarrhea and vomiting. Some diuretics can also cause potassium depletion so that potassium supplement, for instance potassium chloride may be necessary. Effects of potassium depletion include muscular weakness, lethargy and increased sensitivity of the heart to digoxin. Severe depletion can cause death.

Recently it has been shown that potassium depletion may occur in the recovery phase of severe pernicious anemia. Treatment with vitamin B_{12} results in large numbers of red cells, white cells and platelets being produced by the bone marrow, all requiring to be supplied with potassium. So potassium deficiency develops and the patient may die from hypokalemia (Gr. *hypo,* low, +Gr. *kalium,* potassium, +Gr. *haima,* blood). These previously unexplained deaths early in the treatment of severe pernicious anemia can now be prevented by giving the patient potassium chloride by mouth.

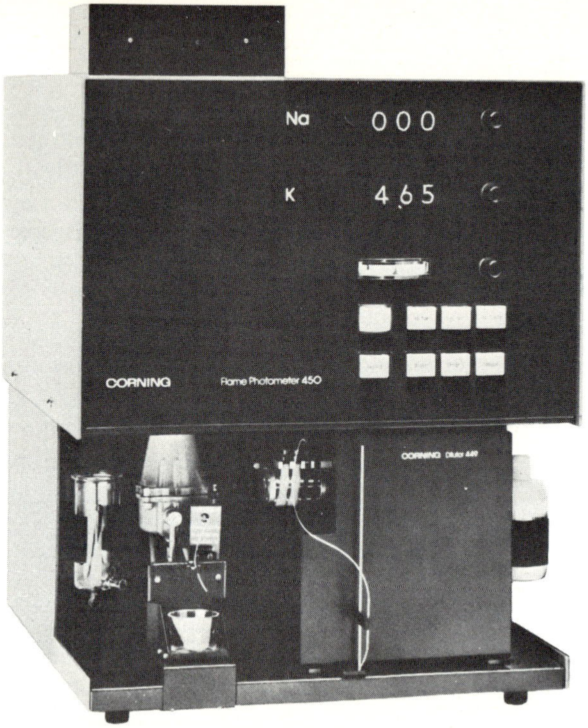

Figure 15.2
Flame photometer, Corning-EEL.

Estimation of sodium and potassium in blood

We have seen how a sodium flame was yellow and a potassium flame lilac. The colors are so characteristic for these elements that they are used in the laboratory for estimating the amounts of these substances in the blood, by means of an instrument called a flame photometer (Fig. 15.2). The instrument is set using standard sodium or potassium solutions. A known dilution of the patient's serum is sprayed into the flame. The intensity of light produced at the appropriate wavelength indicates the concentration of sodium and potassium present. It is very quick and avoids the complicated chemical estimations which used to be necessary. In many hospitals it is now undertaken automatically by a machine such as the autoanalyzer (Autotecnicon Ltd.). It is, however, equally important that the human part of the operation, the collection and labeling of the specimen, be carried out with care and that it be sent to the laboratory without delay. Some of the pitfalls are described at the end of Chapter 13.

16

Chlorine and the Halogen Group

Chlorine was used as a poison gas in the 1914–1918 war causing great disability and loss of life. The same element has also saved many lives and prevented much water-borne disease by making our water supply safe to drink and by disinfecting our swimming pools. The chlorine ions form an essential ingredient of our body fluids in which they are the most abundant anions.

Chlorine is the commonest member of the group of elements called the halogens. Their great chemical reactivity is related to their atomic structure.

ATOMIC STRUCTURE

The five elements in the halogen group are fluorine, F, chlorine, Cl bromine, Br, iodine, I, and astatine, At (Gr. *astatos,* unstable). Astatine is a rare radioactive element made by man, that is so unstable that little is known of its chemistry. Each halogen has seven electrons in its outer (valence) shell

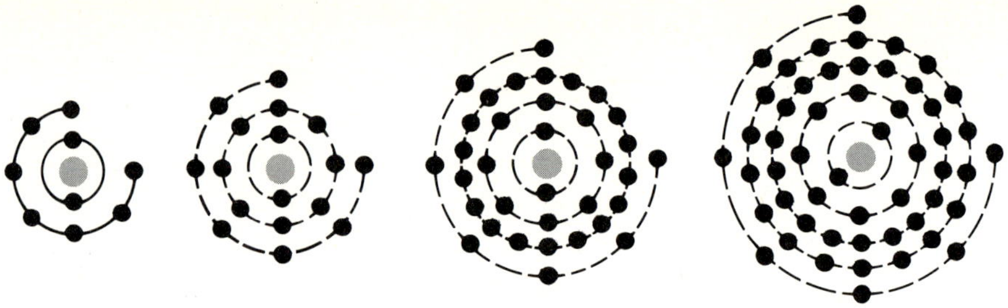

	FLUORINE	CHLORINE	BROMINE	IODINE
Electron configuration:	2, 7	2, 8, 7	2, 8, 18, 7	2, 8, 18, 18, 7
Atomic number:	9	17	35	53
State at room temperature:	pale yellow gas	green gas	dark red liquid	grayish black solid
Melting point:	−223 °C	−102 °C	−7 °C	+114 °C
Boiling point:	−187 °C	−34 °C	+59 °C	+184 °C
Radius of atom (pm):	64	99	114	133

The atoms of the halogens each need one electron to convert the outer (valence) electron shell into a stable octet. By forming a covalent bond between two atoms, as in F_2, Cl_2, Br_2, and I_2 a certain degree of stability is achieved.

But this covalent bond is readily broken in favor of forming an ionic link with a metal. The halogens readily unite with heated metals and are converted into the ions $F^{\ominus}$, $Cl^{\ominus}$, $Br^{\ominus}$, and $I^{\ominus}$. The union often occurs with sufficient vigor to produce a deflagration.

The readiness with which the halogen atoms accept electrons makes them very powerful oxidizing agents. It is this property which, at the right dilution, makes them good antiseptics, able to oxidize bacteria without damaging human tissue. At high concentrations this same property makes them lethal to human tissue, such as chlorine gas acting on the cells of the respiratory tract.

The most reactive of all known elements is the halogen with the smallest atoms, fluorine. Next in reactivity among the halogens is chlorine, then bromine and then iodine. The nucleus of the iodine atom is screened from the outer (valence) electron shell by the four intervening completed electron shells. This diminishes the force available to attract an electron from another element into the gap in iodine's valence shell, so diminishing its degree of reactivity. This screening effect is less in bromine, less still in chlorine and least of all in fluorine where only two electrons intervene between the nucleus and the valence shell. The attraction exerted by the nucleus on an extraneous electron entering the orbit of this fluorine valence shell is powerful, hence its great reactivity.

Iodine with the largest atoms and molecules, I_2 has the highest boiling and melting points, next comes bromine, Br_2, then chlorine, Cl_2, and finally fluorine, F_2 with the lowest boiling and melting points and the smallest atoms in the group. These characteristics are tabulated near the beginning of this section.

FLUORINE, F_2

Fluorine (L. *fluere*, to flow) is a pale yellow corrosive gas which reacts with practically all organic and inorganic substances. Consequently it is highly toxic. Its acid, hydrofluoric acid, HF is so corrosive that it cannot be contained in a glass vessel. In fact, it is used for etching glass. Although fluorine is chemically similar to chlorine it is too reactive to be used as an antiseptic; but fluorides, the salts of hydrofluoric acid, have been found to protect teeth from decay. Even fluorides, which contain the fluoride ion, $F^{\ominus}$, are poisonous unless very diluted; for this reason there has been much controversy over the addition of fluoride to the water supply. Supporters point out that in parts of the world where fluoride occurs naturally in the drinking water at a higher concentration than that recommended for adding artificially, the only ill effect is mottling of tooth enamel, at the same time the teeth being more resistant to decay. Its opponents fear that there will be occasions when a poisonous concentration will occur, perhaps from evaporation in a saucepan or kettle. In places where the drinking water is fluoridated, there have been complaints about the taste of the water; fluoride has also been blamed for gastro-intestinal disorders. But both could be due to other factors. Meanwhile fluoridation of the water supplies continues to be introduced in new areas. So in a few years, it should be possible to form an opinion from personal experience.

Halothane

One fluorine compound of medical importance is the general anesthetic halothane (fluothane, I.C.I.). This is an organic compound containing three different halogens, fluorine, chlorine, and bromine, $CHBrClCF_3$. Structurally it may be represented as

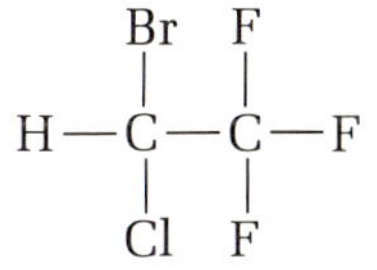

It has proved so satisfactory that it has replaced chloroform $CHCl_3$ as an anesthetic.

CHLORINE, Cl_2

Chlorine (Gr. *Chloros*, greenish yellow) is a greenish yellow gas. Chemically it is sufficiently reactive to combine directly with most elements. Its pungent odor is detectable at 3.5 ppm. At higher concentrations it is a respiratory tract irritant. At 1000 ppm a few deep breaths are fatal and for this reason it was used as a war gas in 1915. It is now widely used in making medicines, antiseptics, insecticides, paints, dye-stuffs, textiles, solvents and many other products. All over the world it is used to make water safe to drink. Its oxidizing action readily destroys bacteria and provides the basis for a number of disinfectants, for example the hypochlorites.

The main acids derived from chlorine are hydrochloric acid and four oxy-acids:

	ACID	FORMULA	SALTS	EXAMPLES
	hydrochloric acid	HCl	chlorides	NaCl
Oxy-acids	hypochlorous acid	HClO	hypochlorites	NaClO
	chlorous acid	$HClO_2$	chlorites	$NaClO_2$
	chloric acid	$HClO_3$	chlorates	$NaClO_3$
	perchloric acid	$HClO_4$	perchlorates	$NaClO_4$

Hydrochloric acid HCl

Hydrochloric acid is a typical strong mineral acid with the characteristics described in Chapter 13. It is an aqueous solution of hydrogen chloride that is completely dissociated in water into separate hydrogen and chloride ions, $H^{\oplus}$ and $Cl^{\ominus}$. The high concentration of $H^{\oplus}$ ions gives it the properties of a strong acid able to digest living tissue. It is, therefore, surprising that the glands of the stomach can produce this strong acid in high concentration without damage to themselves or to the other cells lining the stomach wall. That this secretion has mineral acid properties was demonstrated by an army surgeon, Dr. William Beaumont, in 1819. He studied a patient, Alexis St. Martin, a French Canadian hunter, with a stomach opening on to the abdominal wall caused accidently by a musket wound. He found, for example, that the hydrochloric acid produced in this stomach could digest a living frog's leg.

Hydrochloric acid, reacts with alkalis to form salts called chlorides (see Chapter 15). Its USP category is "pharmaceutic aid (acidifying agent)."

Sodium chloride, NaCl

See page 102.

Chloride in the human body

The chloride ion, $Cl^{\ominus}$ is the most abundant anion in our extracellular fluid, the normal blood level being 96 to 106 mmol/liter. In the healthy body, this level is very accurately maintained. The normal daily intake is 10 to 15 g (approximately 200 to 300 mmol) in the form of sodium chloride and is balanced by a similar loss in urine, sweat and feces. The mechanism by which the balance is achieved is essentially the same as for sodium, although the distribution of sodium and chloride ions in the body is not identical. In certain conditions chloride loss exceeds intake, resulting in chloride depletion (hypochloremia). This occurs in excessive sweating, vomiting or diarrhea, also from excessive urinary loss as in diabetes mellitus, Addison's disease and certain infections, particularly pneumonia and pulmonary tuberculosis. With the exception of Addison's disease the body responds to the low serum chloride by reducing or ceasing chloride excretion in the urine. So the finding of a low chloride concentration in the urine indicates hypochloremia. This is more directly demonstrated by serum electrolyte estimation.

The remedy for chloride depletion is to administer chloride, usually in the form of sodium chloride solution, either by mouth or by intravenous saline (sodium chloride) infusion. A raised chloride (hyperchloremia) can also occur namely when excretion is impaired, as in a urinary tract obstruction, severe kidney disease and cardiac failure.

Hypochlorous acid, HClO

Hypochlorous acid differs greatly from hydrochloric acid. It is a weak acid and so in solution it dissociates only slightly into ions.

$$\underset{(aq)}{HClO} \rightleftharpoons \underset{(aq)}{H^{\oplus}} + \underset{\substack{(aq)\\ \text{HYPOCHLORITE} \\ \text{ION}}}{ClO^{\ominus}}$$

Hypochlorous acid solution is pale yellow. It is only stable if kept cool and in the dark. On heating or on exposure to light the concentrated solution yields oxygen and chlorine.

$$\underset{(aq)}{4HClO} = \underset{(g)}{O_2} + \underset{(g)}{2Cl_2} + \underset{(l)}{2H_2O}$$

The salts of hypochlorous acid are called hypochlorites. Since they are potent oxidizing agents they can be used for bleaching such as bleaching powder, which contains calcium hypochlorite, $Ca(ClO)_2$ and as antiseptics such as sodium hypochlorite, NaClO.

Sodium hypochlorite, NaClO

Sodium hypochlorite is a powerful oxidizing agent. In its pure form it is extremely unstable, readily releasing oxygen and chlorine. Its powerful oxidizing activity makes it an effective bleaching agent and antiseptic.

Milton's fluid is a 1% aqueous solution of sodium hypochlorite, also containing 16.5% sodium chloride. This fluid, at a strength of 12.5 cm^3/liter (1 in 80), is used to disinfect babies' bottles and nipples. To ensure sterilization they must be completely immersed in a freshly prepared solution for $1\frac{1}{2}$ to 3 hours. Articles to be sterilized must first be thoroughly cleaned. If there is any milk or other protein present, the chlorine combines with it and is no longer available for disinfection. Sodium hypochlorite is also the essential ingredient of Dakin's solution and several proprietary preparations, including Zonite (Dunbar Labs.), Chlorpactin (Guardian Chemical Corp.) and Chloros (I.C.I.).

Chloroform, $CHCl_3$

Chloroform has a molecule like that of methane, CH_4 with three of the four hydrogen atoms replaced by chlorine.

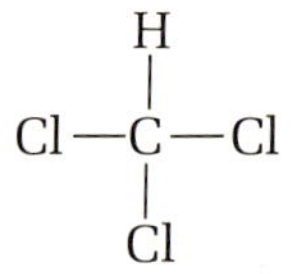

Its three dimensional structure is described in Chapter 21.

Chloroform is a liquid with a low boiling point 61°C and a sweetish smell. It is not flammable but its vapor can burn. At one time it was widely used as a general anaesthetic but it can cause severe or even fatal liver damage and so it is no longer used for this purpose. Halothane, whose chemical structure is slightly similar has taken its place. Chloroform is now mainly used as a preservative in liquid medicines. Its presence has little effect on an adult but does affect the susceptible nervous system of babies. So medicines for the newborn are made up without chloroform preservatives; this means that they do not keep and should be discarded after a week.

Chloroform exposed to air and sunlight undergoes gradual oxidation to poisonous phosgene.

$$2CHCl_3 + O_2 \xrightarrow{\text{sunlight}} \underset{\text{PHOSGENE}}{2COCl_2} + 2HCl$$

So chloroform is stored in brown bottles with 0.6 to 1.0% alcohol which destroys any phosgene which may be formed during storage.

Carbon tetrachloride, CCl_4

Carbon tetrachloride has a molecule like that of methane with all four of the hydrogen atoms replaced by chlorine.

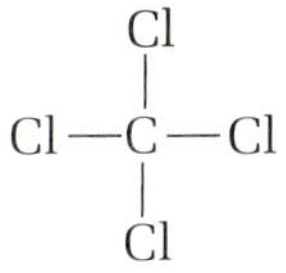

Its three-dimensional structure is illustrated in Chapter 21. Like chloroform it is a liquid with a low boiling point (77°C) and a sweetish smell but, unlike chloroform, it is noncombustible. For this reason it is used in some fire-extinguishers, for example, Pyrene.

When sprayed on to a fire it vaporizes and the heavy vapor settles over the flames like a blanket, smothering the fire by excluding oxygen. Some phosgene is formed and so it should not be used in a confined space, particularly since carbon tetrachloride is an even more potent liver toxin than chloroform. Carbon tetrachloride is a very effective solvent for grease and oil. It has been widely used as a cleaning fluid but inhalation of its vapor quite rapidly produces fatty degeneration of the liver.

BROMINE, Br_2

Bromine (Gr. *bromos*, stench) is the only nonmetallic element that exists in liquid form at ordinary temperatures. It is reddish brown in color. At room temperature it gives off a red vapor with a strong disagreeable odor, resembling chlorine, and is very irritating to the eyes and throat. If spilled on the skin it causes painful sores.

Bromine readily unites with many elements to form bromides. At one time bromides were widely prescribed as sedatives, a popular mixture being potassium bromide and chloral. They have now been superceded by a large variety of proprietary preparations. Bromides are still much used in photography and bromine compounds are incorporated into many medical preparations.

IODINE, I_2

Iodine (Gr. *iodes,* violet) is a grayish-black lustrous solid. When slightly warmed it gives off a blue-violet vapor with an irritating odor. Iodine has considerable medical importance. It is an essential ingredient of our diet although only very small amounts are needed, about 0.15 mg/day. A trace of iodine (0.001%) is present in the sea. Its concentration is somewhat higher in fish, fish liver oil and seaweed. Ash from seaweed, called kelp, contains 1% iodine. At one time goiter due to iodine deficiency was not uncommon in Switzerland whose borders are well away from the sea. Since 1922 iodized salt, containing 1 in 200,000 iodine in the form of potassium iodide, KI has been available and goiter has now become rare in Switzerland. No mother taking such salt during the last five months of pregnancy has given birth to a child with goiter. Iodine in the form of iodine tincture (USP) or weak iodine solution (BP) has long been used as an antiseptic and is still considered as one of the more effective skin disinfectants. Weak iodine solution (BP) is a 2.5% solution of iodine (with 2.5% potassium iodide and 2.5% water) in ethyl alcohol. Iodine tincture (USP) is similar but contains 2% iodine. Both the alcohol and the iodine have an antiseptic action. Like the other halogens, iodine kills bacteria by oxidizing them. This effect is diminished by the presence of organic matter. So the site to be treated should be previously cleaned. Iodine must not be used to cleanse the skin prior to vaccination as it kills vaccinia virus. Its use must also be avoided for patients who have a sensitivity to iodine.

Sodium Iodide, NaI and potassium iodide, KI

See Chapter 15.

Iodine in the body

The iodine from the diet is taken up by the thyroid gland to make thyroxine which, combined with protein, is the main ingredient of the thyroid hormone. Each molecule of thyroxine T_4 (tetra-iodothyronine) contains four atoms of iodine.

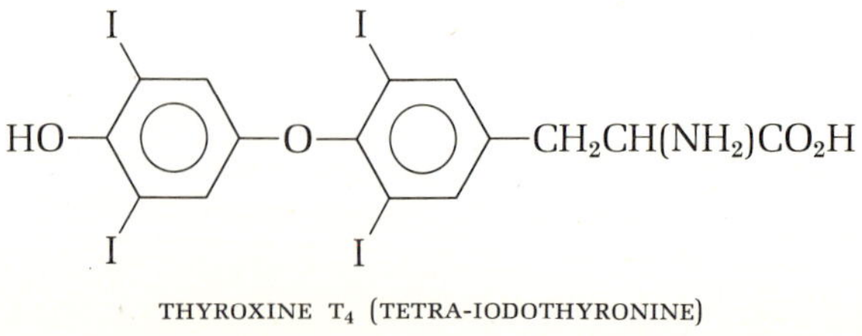

THYROXINE T_4 (TETRA-IODOTHYRONINE)

The other thyroid hormone, called T_3 or tri-iodothyronine, has a very similar structure but contains only three atoms of iodine.

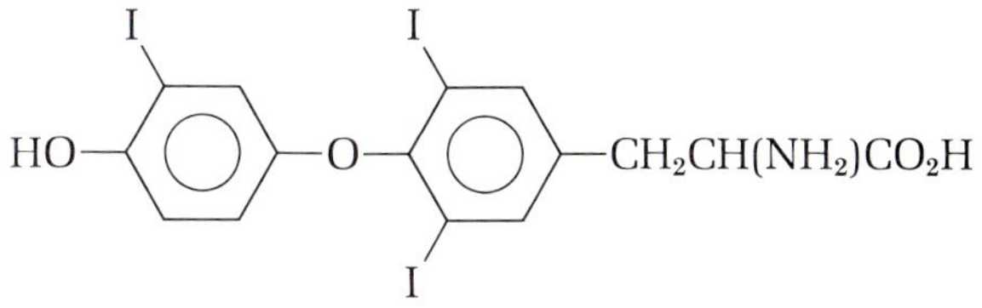

T_3 (TRI-IODOTHYRONINE)

Although the formula of thyroxine appears rather complicated it shows a feature which is characteristic of many of the chemical compounds in the body. It has a definite shape: the four iodine atoms of T_4 protrude at the sides, like arms and legs; there is a hydroxyl OH group at one end, like a head, with a fairly long chain at the other end like a tail; in the middle is an oxygen atom, like a wasp-waist between the two hexagons. Each hexagon represents six carbon atoms arranged in a ring called a benzene ring.

There is no need to memorize the formula of thyroxine, only to note that it has a recognizable shape. For the shape of a molecule often plays an important part in its biological activity. Both thyroxine T_4 and T_3 have marked biological activity. Without them every cell in the body tends to function less well than normal. A baby born without a functioning thyroid gland becomes a cretin. Both mental and physical development are retarded, and unless treated the baby will become a stunted imbecile.

Too much thyroid secretion (hyperthyroidism or thyrotoxicosis) causes the rate of metabolism (chemical reactions in cells) to be excessive. The heart beats too rapidly and the nervous system becomes overexcitable. Severe overactivity or underactivity of the thyroid gland is usually obvious clinically. Minor abnormalities can be almost as serious over a long period of time but are much more difficult to detect. The estimation of protein bound iodine, T_4 and T_3 in the blood helps to demonstrate such abnormalities. It is also helpful to measure the amount of T_3 which can be taken up by the blood proteins (T_3 uptake).

Protein bound iodine, PBI

One way in which thyroid activity can be assessed is by measuring the amount of iodine which is bound to protein in the blood. The normal PBI level is 4–8 $\mu g/100\ cm^3$. It is raised when thyroid activity is excessive (hyperthyroidism) and reduced when it is deficient (hypothyroidism) as in cretinism and myxedema. Unfortunately, false results can occur from only slight contamination by iodine, either in the form of the element, as in iodine tincture, or in a compound such as lipiodol. The latter is used as a contrast medium in the X ray department and sufficient iodine may remain in a patient's serum to invalidate the test for 6 to 12 months.

Thyroxine T_4

Thyroxine T_4 (Tetra-iodothyronine) accounts for 90% of the protein bound iodine but its estimation is much less liable to be affected by iodine contamination. The normal serum thyroxine is 50 to 140 mmol/liter (4 to 11 μg/100 cm^3). It is sometimes reported as thyroxine iodine with a normal value of 2.6 to 7.2 μg/100 cm^3 (50 to 140 mmol/liter). Its estimation involves the use of radio-iodine (see below). Interpretation of results is similar to that for PBI.

T_3 Tri-iodothyronine uptake

The measurement of serum thyroxine T_4 can be affected by various influences such as by pregnancy and the contraceptive pill. So it is necessary to have a means of checking it. Measurement of the T_3 uptake by a sample of serum provides such a check. Radio-iodine (see below) is incorporated into the T_3 so that the amount taken up can be conveniently measured.

The hormone thyroxine T_4 is bound to serum proteins which have a limited number of binding sites on them. The more sites which are occupied by thyroxine T_4 the fewer sites there are left to combine with the radioactive T_3 used in the test. Normally 25–36% of the sites are unoccupied and therefore free to combine with T_3. The increased amount of thyroxine in thyrotoxicosis means that more sites are occupied. So the number of free sites is reduced to well below 25%. Conversely in cretinism and myxedema the number of free binding sites is increased considerably above 36%.

Free thyroxine index

The free thyroxine index provides a measure of the actual biological activity of thyroxine T_4, compensating for alterations due to pregnancy and oral contraceptives. It is calculated from the product of the T_4 level and the T_3 uptake. The normal range is 2.2 to 7.1. It is raised in thyrotoxicosis and reduced in myxedema and cretinism.

T_3 Tri-iodothyronine estimation

T_3 has four times the biological activity of T_4. Normally the ratio of T_4 to T_3 in blood is 50:1 but in some cases of thyrotoxicosis it is mainly the T_3 that is raised. This is best demonstrated by estimating the T_3 in blood; methods have recently been developed to do this.

Radio-iodine

Radio-iodine is a radioactive isotope of iodine. It has to be prepared artificially by bombarding iodine with subatomic particles. The atoms produced have an altered atomic weight and are unstable. Their radioactivity is the result of this instability. The radiation emitted is readily detected and measured by means of an instrument such as a Geiger counter.

Radio-iodine has a number of medical applications. The isotope most frequently used is ^{131}I which has a half-life of just over eight days. This is long enough to be usable but short enough not to be a radiation hazard if properly used. The half-life is the time taken for half of a radioactive element to disintegrate. Recently ^{132}I with a half-life of 2 to 3 hours has become available.

Medical uses of radio-iodine include the estimation of Thyroxine T_4 and of T_3 uptake, already described, also more recently of serum T_3 itself which may be of more biological importance than was previously thought. Another application is the radio-iodine test of thyroid activity in which the patient, after fasting for at least two hours, is given a measured dose of radio-iodine to drink; urine is collected at timed intervals over the next 48 hours. In thyrotoxicosis more iodine than usual is taken up by the thyroid so that the amount excreted in the urine is reduced. The reverse occurs in myxedema. In selected cases the uptake by the thyroid can be measured directly by a suitable counter. Radio-iodine has also been used for the detection of deposits of thyroid cancer outside the thyroid gland and, occasionally, for the treatment of thyrotoxicosis and of thyroid cancer.

17

Magnesium and Calcium

Magnesium and calcium are chemically related, both belonging to the group known as the alkaline earth metals. All the members of this group are of medical importance:

ELEMENT (SYMBOL)	ATOMIC NUMBER	ELECTRON CONFIGURATION	MEDICAL SIGNIFICANCE
Beryllium (Be)	4	2, 2	Beryllium compounds cause berylliosis, a disease affecting the lungs and soft tissues.
Magnesium (Mg)	12	2, 8, 2	Essential for life, being the second most abundant intracellular metal after potassium. Affects contract-

ELEMENT (SYMBOL)	ATOMIC NUMBER	ELECTRON CONFIGURATION	MEDICAL SIGNIFICANCE
			ility of muscle and excitability of nervous tissue.
Calcium (Ca)	20	2, 8, 8, 2	Gives bones their rigidity (and X-ray opacity, Fig. 17.1). Essential for blood coagulation. Affects muscle contractility and nervous excitability.
Strontium (Sr)	38	2, 8, 18, 8, 2	Strontium is treated by the body like calcium and deposited in bone, as is radio-strontium which constitutes a long-acting radiation hazard.

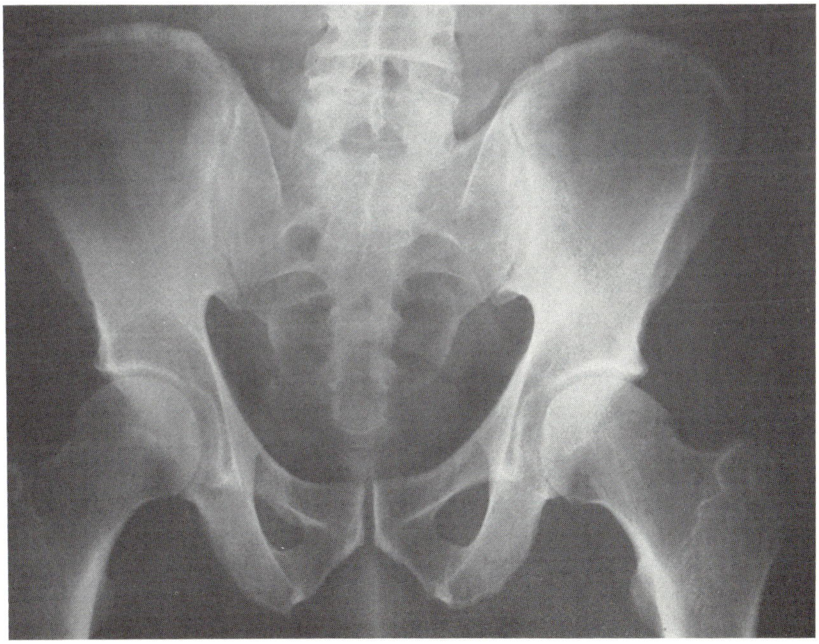

Figure 17.1
X Ray of pelvis; calcium in bones makes skeletal structure visible. (Courtesy Dr. J. Hinde.)

ELEMENT (SYMBOL)	ATOMIC NUMBER	ELECTRON CONFIGURATION	MEDICAL SIGNIFICANCE
Barium (Ba)	56	2, 8, 18, 18, 8, 2	Barium salts are opaque to X rays. If a patient swallows a barium meal, X-ray examinations will show up abnormalities of the alimentary tract (Fig. 17.2).
Radium (Ra)	88	2, 8, 18, 32, 18, 8, 2	Radium is radioactive and used in the treatment of cancer.

ATOMIC STRUCTURE

As the above table indicates, all the alkaline earth metals have a similar atomic structure in that each has two electrons in the outer valency shell. The structures of magnesium and calcium can be represented thus:

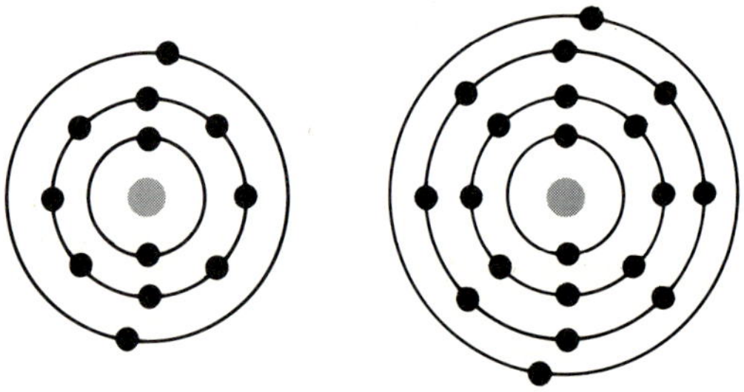

	MAGNESIUM	CALCIUM
Atomic number	12	20
Electron configuration	2, 8, 2	2, 8, 8, 2

It can be seen that their atomic structure resembles that of the alkali metals sodium and potassium (see Chapter 15) with which, chemically, they have much in common. Having two electrons in the outer shell they have a valency of two, as opposed to the one of the alkali metals. Magnesium and calcium have a harder consistency than sodium and potassium and their hydroxides have a weaker alkalinity than those of the alkali metals. Magnesium hydroxide, or Milk of Magnesia, for example, is widely used as an antacid for stomach complaints; in contrast sodium

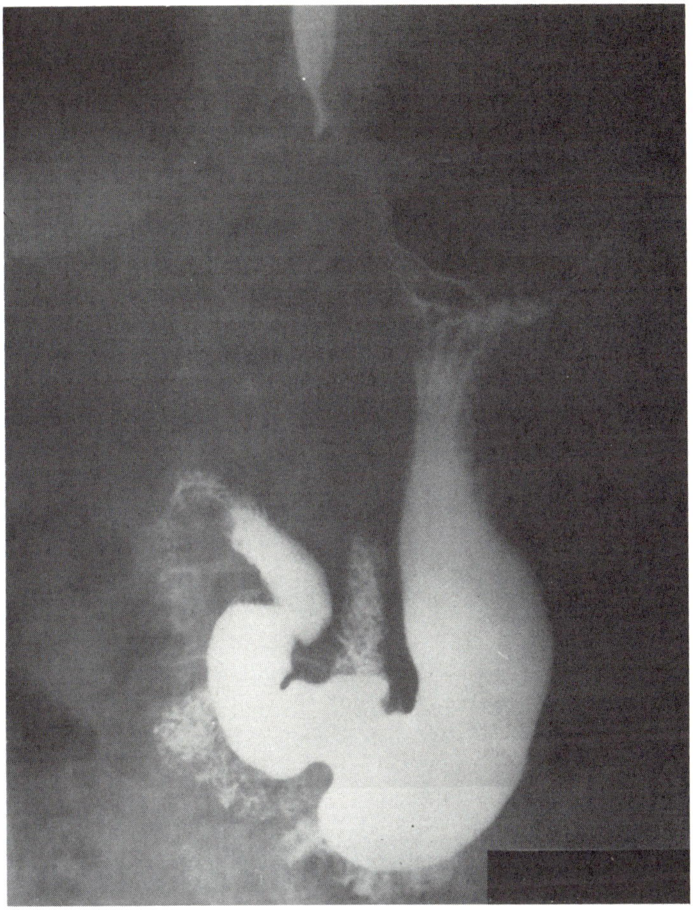

Figure 17.2
Barium meal showing stomach outline on the X ray. Barium can also be seen in the esophagus, duodenum and upper jejunum. (Courtesy Dr. J. Hinde.)

hydroxide is so alkaline that it is corrosive and can destroy the stomach wall.

Calcium differs from magnesium in having an extra electron shell with its octet of electrons intervening between the nucleus and the two outer valence electrons. The effect of this is partially to screen these outer electrons from the attracting force of the nucleus. As a result calcium releases these two electrons more readily than magnesium to form the simple cation $Ca^{2\oplus}$. Part of the calcium in our blood is present in this ionic form. If there is insufficient ionized calcium in our blood, our muscles go into involuntary spasm (tetany). Insufficient magnesium in the blood also produces tetany.

ATOMIC ABSORPTION

We have seen how atoms have a structure which is specific for each individual element. As a result the atoms absorb light of a wavelength which is specific for each individual element. If a solution of a substance containing magnesium or calcium is aspirated into a flame, an atomic vapor is formed. When light of the appropriate wavelength is passed through the flame, part of it is absorbed. The amount of light absorbed is proportional to the density of the relevant atoms in the flame. This principle, known as atomic absorption spectroscopy, is now used in hospital laboratories for measuring the amount of magnesium and calcium in blood plasma (Fig. 17.3).

MAGNESIUM

Magnesium (from Magnesia, in Thessaly) is a light, silvery-white metal. It is fairly tough and is one-third lighter than aluminum, with which it is used in alloys for constructing aircraft, space rockets and missiles. It was first isolated by Davy in 1808 although its existence as an element had previously been postulated by Black in 1755.

Magnesium is the eighth most abundant element by weight in the earth's crust, where it occurs in combined form as magnesite, dolomite,

Figure 17.3
Atomic absorption spectrophotometer, Corning-EEL.

talc, asbestos and other minerals. The metal burns in air with a dazzling white flame to produce magnesium oxide.

$$\underset{(s)}{2Mg} + \underset{(g)}{O_2} = \underset{(s)}{2MgO}$$

This property is used in flashlight photography, flares and fireworks.

Medical applications

A number of magnesium compounds are used in medicine. Probably the best known is *magnesium hydroxide,* $Mg(OH)_2$ (or "Milk of Magnesia) used as a mild antacid usually in the form of a suspension in water. This suspension consists of particles, each of which is an aggregation of magnesium and hydroxyl ions. These particles remain suspended in watery solution for only a limited time, eventually sedimenting to the bottom of the bottle. So it is important to shake the bottle well before use. Its USP category is "antacid; cathartic." Another antacid much used in the treatment of gastric ulcers is *magnesium trisilicate,* $2MgO \cdot 3SiO_2$(with water of crystallization). It neutralizes the excess acid rather slowly and it is long acting. It does not stimulate the stomach to produce more acid and there is no danger of it causing alkalosis, both of which occur with sodium bicarbonate. Its category is "antacid" only.

Another magnesium compound of medical importance is *magnesium sulfate,* $MgSO_4$ which usually occurs combined with "water of crystallization" as $MgSO_4 \cdot 7H_2O$ as in crystals of Epsom salts. For simplicity it may be imagined to consist of magnesium cations and sulfate anions

$Mg^{2\oplus}$ $SO_4^{2\ominus}$

In fact, the magnesium ions probably combine with the water molecules to form complex cations which have to be visualized in three dimensions, like this:

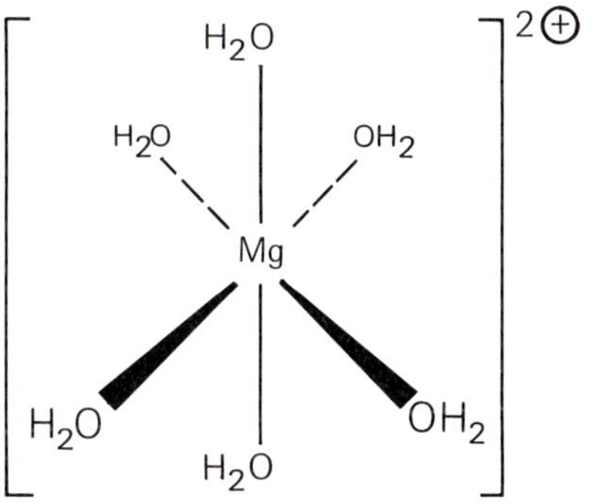

The use of magnesium sulfate as an osmotic purgative and poultice has already been described (Chapter 14). Its USP category is "anticonvulsant; cathartic."

Magnesium in the body

Nearly three-quarters of the magnesium in the body is present in the skeleton, combined with calcium and phosphorus to form the complex salts which give bone its rigidity. Most of the remainder is situated within the body cells where magnesium is the next most abundant cation to potassium. Several enzymes require magnesium as a cofactor. The normal plasma magnesium is 0.6 to 1 mmol/liter (1.5 to 2.6 mg/100 cm^3). A low serum magnesium causes neuromuscular irritability and tetany, similar to that caused by a low serum calcium. Raised serum magnesium can produce coma and, surprisingly, this effect is counteracted by giving calcium in the form of calcium gluconate.

Magnesium is essential in the diet although the daily need is not known. A diet deficient in magnesium must be very rare, but magnesium deficiency can occur as a result of severe diarrhea, kidney failure or infected burns.

CALCIUM

Calcium (L. *calx*, lime) is a rather hard, silvery metal. Its compounds give rigidity to bone, teeth and shells. Although lime was prepared in the first century A.D. by the Romans, who called it calx, the metal was not isolated until 1808 by Davy. Calcium is the fifth most abundant element in the earth's crust. It occurs as limestone (calcium carbonate, $CaCO_3$), gypsum (calcium sulfate, $CaSO_4 \cdot 2H_2O$), and other minerals. Chalk, derived from bygone shellfish, is also calcium carbonate.

Calcium metal burns with a yellow–red flame in air mainly to form calcium nitride with some oxide

$$3Ca + N_2 = \underset{\text{CALCIUM NITRIDE}}{Ca_3N_2}$$

$$2Ca + O_2 = \underset{\text{CALCIUM OXIDE}}{2CaO}$$

The yellow–red color of the flame is a characteristic of the metal and is used for estimating the amount of calcium in the blood by means of a flame photometer. The metal also combines readily with chlorine

$$Ca + Cl_2 = CaCl_2$$

CALCIUM CHLORIDE

with sulfur

$$Ca + S = CaS$$

CALCIUM SULFIDE

and with hydrogen

$$Ca + H_2 = CaH_2$$

CALCIUM HYDRIDE

Quick-lime

Quick-lime is calcium oxide, CaO. It is usually made by heating calcium carbonate as limestone in a lime kiln (Fig. 17.4). Heat splits carbon dioxide off from the calcium carbonate, leaving calcium oxide.

$$\underset{(s)}{CaCO_3} = \underset{(s)}{CaO} + \underset{(g)}{CO_2}$$

CALCIUM CARBONATE (LIMESTONE) — CALCIUM OXIDE (QUICK-LIME)

Calcium oxide provides the basis of many products made by the chemical industry.

Slaked lime

Slaked lime is calcium hydroxide $Ca(OH)_2$. It is made by spraying water on to quick-lime.

$$\underset{(s)}{CaO} + \underset{(l)}{H_2O} = \underset{(s)}{Ca(OH)_2}$$

It has many industrial uses, such as in mortar, cement and the manufacture of bleaching powder which contains calcium hypochlorite, $Ca(ClO)_2$. Limewater is calcium hydroxide dissolved in water. It is a colorless solution. If carbon dioxide is passed into limewater it turns cloudy.

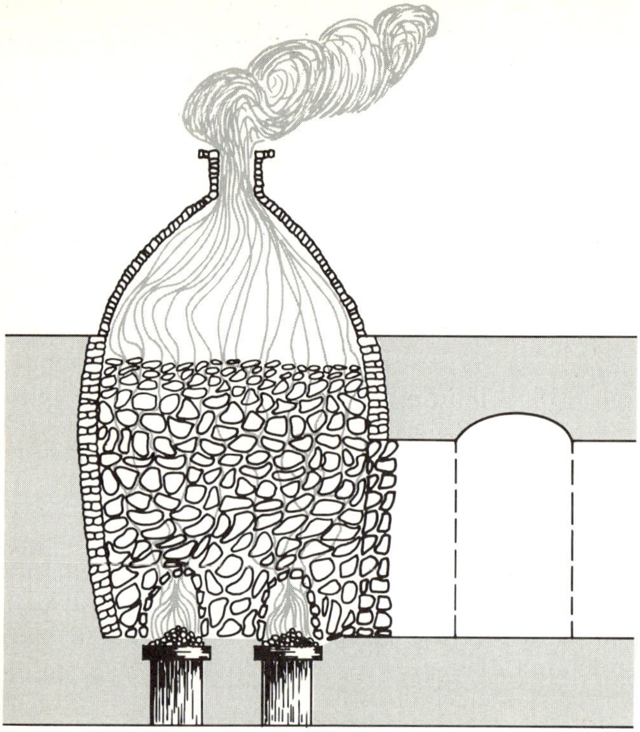

Figure 17.4
Lime kiln.

$$Ca(OH)_2 + CO_2 = CaCO_3 + H_2O$$

(aq) (g) (s) (l)
CALCIUM HYDROXIDE — CALCIUM CARBONATE

The cloudiness is due to the precipitate of calcium carbonate. The readiness with which limewater turns cloudy provides a sensitive means of detecting carbon dioxide, for example in expired air.

If carbon dioxide continues to be bubbled through the cloudy limewater it eventually clears again.

$$CaCO_3 + H_2O + CO_2 = Ca(HCO_3)_2$$

(s) (l) (g) (aq)
CALCIUM BICARBONATE

The insoluble calcium carbonate is converted into soluble calcium bicarbonate.

Calcium carbonate

Calcium carbonate, $CaCO_3$ is the essential constituent of limestone and chalk. In purified form it is used medically as an antacid in the treatment of excessive gastric acidity.

$$CaCO_3 + 2HCl = CaCl_2 + H_2O + CO_2$$

It is also added to flour to provide adequate calcium in the diet. Its USP category is "antacid."

Calcium chloride

Calcium chloride, $CaCl_2$ is a soluble calcium salt. It is used in the treatment of children suffering from calcium deficiency, given by mouth. In the past it was used by injection for the treatment of tetany, in adults and children, due to low blood calcium. However, sometimes it has the unfortunate effect of causing the heart to contract and stop ("calcium rigor"). For this reason, for intravenous use calcium chloride has now been replaced by calcium gluconate which does not produce such a high concentration of calcium ions. Its USP category is "calcium replenisher."

Plaster of Paris

Plaster of Paris is a form of calcium sulfate known as the hemihydrate $2CaSO_4{\cdot}H_2O$ (sometimes written $CaSO_4{\cdot}\frac{1}{2}H_2O$). When plaster of Paris is mixed with water it rapidly sets and hardens.

$$\underset{\text{PLASTER OF PARIS}}{2CaSO_4 \cdot H_2O} + 3H_2O \longrightarrow \underset{\text{GYPSUM}}{2(CaSO_4 \cdot 2H_2O)}$$

In the process of being converted into gypsum the plaster of Paris expands slightly.

Calcium and the human body

About 99% of the calcium in our body is in our bones and gives them rigidity. The remaining 1% has a number of very important functions. In the extracellular fluid the presence of calcium ions is critical in preventing excessive excitability of the nerves and muscle; it also plays a vital role in blood clotting. Intracellularly, calcium takes part in a number of enzyme systems, including those responsible for conducting nerve impulses and for causing muscular contraction.

A healthy adult requires about 40 mmol (1.6 g) of calcium daily. Up to 125 mmol (5 g) daily are needed during pregnancy and lactation and also for growing children. Cow's milk contains 15 mmol (0.6 g)/pint, mainly as calcium phosphate; cheese contains 20 mmol (0.8 g) calcium/100 g. Flour has added calcium carbonate so that bread can be an important dietetic source. Unfortunately, flour of high extraction (85% or more) contains phytic acid which interferes with calcium absorption. In order to absorb calcium it is vital to have adequate vitamin D in the diet.

Absorption of calcium occurs in the small intestine. It is favored by the relatively high hydrogen ion concentration (low pH) which is found in the upper part of the small intestine due to the carry-over of acid from the stomach. In an acid solution the calcium phosphate formed is soluble whereas in an alkaline solution the phosphate and carbonate produced are insoluble, which results in less absorption in the lower small intestine.

Another important factor is the rate of passage of the intestinal contents. In intestinal hurry, as in coeliac disease, there is insufficient time for absorption of either vitamin D or calcium, resulting in rickets or osteomalacea (adult rickets), with softening and distortion of the bones.

The normal blood calcium is 2.25 to 2.75 mmol/l. (It is still sometimes expressed as 9.0 to 11.0 mg/100 ml.). This blood level is mainly controlled by the parathyroid glands which are situated just behind and below the thyroid gland. The hormone secreted by these glands, called parathormone, mobilizes calcium from the bones thus increasing the blood level; it also facilitates the absorption of calcium from the small intestine. Accidental removal of parathyroid glands has occasionally occurred during thyroidectomy (surgical removal of the thyroid gland). It produces a low serum calcium and if the ionized calcium is sufficiently reduced, it causes tetany, characterized by increased nervous irritability and associated muscle spasms. Sometimes a parathyroid gland tumor occurs which secretes too much parathormone; too much calcium is then mobilized, with a corresponding rise in serum calcium and calcium excretion in the urine, resulting in the bones becoming unduly decalcified and softened (osteitis fibrosa cystica or hyperparathyroidism).

There is another hormone called calcitonin (or thyrocalcitonin) that is secreted by the thyroid gland. The effect of this is the opposite of parathormone in that it promotes the deposition of calcium in the bones from the blood.

The bones consist of fibrous connective tissue (collagen) which is impregnated with calcium salts, mainly calcium phosphate. The process by which the calcium and the phosphate ions in solution in the blood are converted into the hard structure of bone involves the bone cells, called osteoblasts, and an enzyme called phosphatase, but the details are not yet fully understood. The calcium salts in bone are arranged in a crystalline lattice which presents a very large surface area to the interstitial fluid. This means that not only can calcium and sodium be very readily mobilized from bone but also that other substances such as strontium (including radioactive strontium) and lead can readily be laid down in bone.

18 Nitrogen and Phosphorus

Both nitrogen and phosphorus are important elements in medical chemistry. Nitrogen is something of a paradox. It is an inert gas, remaining uncombined in the air, and yet it forms very large numbers of compounds, including proteins which are the essence of life. Phosphorus is a far more reactive element. Like nitrogen it also forms numerous compounds, including the complex phosphoproteins in our brains and the inorganic phosphates which help to buffer the pH in our plasma.

ATOMIC STRUCTURE

Nitrogen and phosphorus are group 5 elements, that is, they occur in group 5 of the periodic table. These elements are characterized by the presence of five electrons in the outer valence shell.

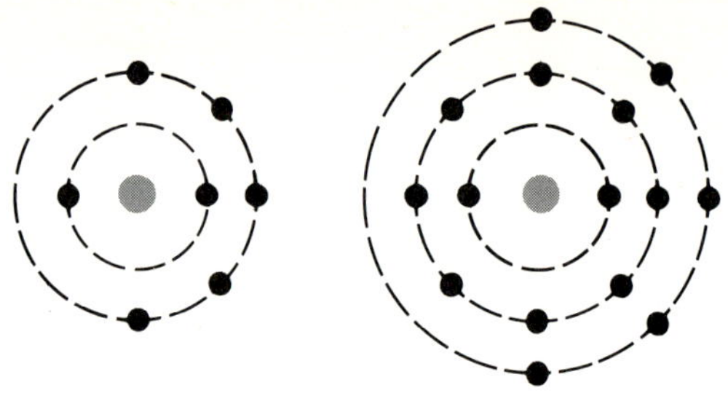

	NITROGEN	PHOSPHORUS
Electron configuration	2, 5	2, 8, 5
Atomic number	7	15

It will be remembered that eight electrons in the outer shell is the stable (rare gas) configuration. Each nitrogen and phosphorus atom forms three covalent bonds with other atoms in order to acquire a stable octet. Ammonia, NH_3 and phosphine, PH_3 illustrate this tervalency of nitrogen and phosphorus. The stereostructures of these molecules may be considered as tetrahedral with the group 5 atom at the center, a hydrogen atom at three of the apexes, and a lone pair of electrons at the fourth apex.

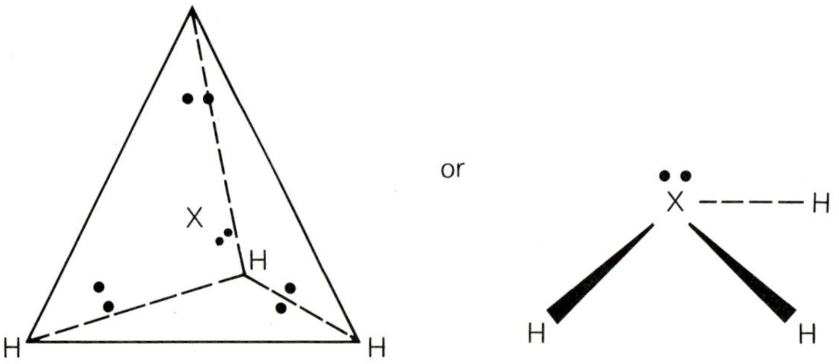

X = N or P

If the lone pair is shared with a hydrogen ion, $H^{\oplus}$, the ammonia molecule is converted into the ammonium ion, $NH_4^{\oplus}$, and the phosphine molecule into the Phosphonium ion, $PH_4^{\oplus}$. A fourth hydrogen atom then occupies the site previously held by the lone pair.

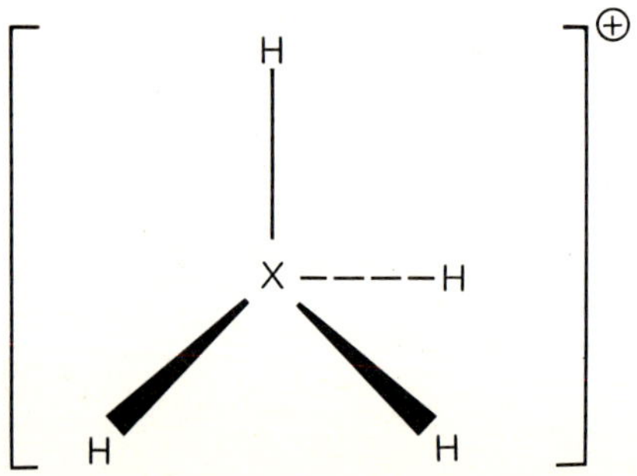

The NH_4 and PH_4 groupings can only exist in the form of ions as in ammonium chloride, $NH_4^{\oplus}Cl^{\ominus}$, and phosphonium iodide, $PH_4^{\oplus}I^{\ominus}$. An atom of phosphorus, being larger than a nitrogen atom, can also accommodate ten electrons in its outer shell and is therefore able to form five covalent bonds with other atoms, as it does in phosphorus pentachloride, PCl_5.

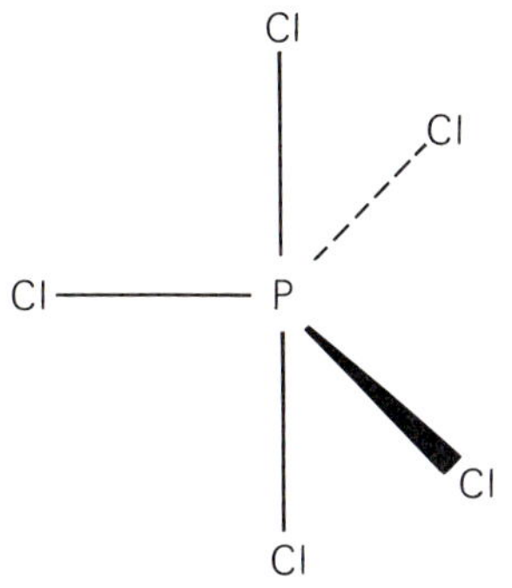

The structure of this molecule may be compared to a three-bladed propellor on a vertical spindle.

NITROGEN

Nitrogen (L. *nitrum,* a salt + L. *genes,* forming) is a colorless, odorless, and tasteless gas, that neither burns nor, in general, supports combustion. It forms 78% (approximately $\frac{4}{5}$) of the air by volume. It was discovered by Daniel Rutherford in 1772, although at that time several other chemists were studying the nature of air from which the oxygen had been removed, calling it "burnt air" or "dephlogisticated air." Nitrogen appeared so inactive chemically that Lavoisier called it "azote" (without life). Yet it is an essential element in all proteins, the substances on which all life depends. Nitrogen compounds provide food, fertilizers, poisons and explosives. Liquid nitrogen, which boils at −196°C and freezes at −210°C, is clear, colorless and odorless, resembling water. It is used as a refrigerant; for example in preserving human red blood cells, which may be stored indefinitely in this deep frozen state. Liquid nitrogen is used extensively in missile work for purging compounds and insulating space chambers. In the oilfield nitrogen gas produced from liquid nitrogen provides the pressure for forcing up the crude petroleum.

Nitrogen, like oxygen, is prepared commercially by fractional distillation of liquid air. Its cost varies from a few cents to over $3/100 ft³, depending on its purity. In the laboratory nitrogen can be prepared by heating an aqueous solution of ammonium nitrite, NH_4NO_2.

$$\underset{(aq)}{NH_4^{\oplus}} + \underset{(aq)}{NO_2^{\ominus}} = \underset{(g)}{N_2} + \underset{(g)}{2H_2O}$$

Since ammonium nitrite is not usually available as such in the laboratory its component ions $NH_4^{\oplus}$ and $NO_2^{\ominus}$ can be obtained in solution by mixing equimolar proportions of ammonium chloride, $NH_4^{\oplus}Cl^{\ominus}$ and sodium nitrite, $Na^{\oplus}NO_2^{\ominus}$ in water.

Although nitrogen is, in general, an inert element it will combine directly with magnesium, calcium or lithium when heated, to form their nitrides. If a mixture of nitrogen and oxygen is subjected to electric sparks it forms first nitric oxide, NO and then nitrogen dioxide, NO_2. At ordinary temperatures and pressures nitrogen scarcely combines with hydrogen at all, but when they are heated together under pressure with a catalyst a 60% yield of ammonia is produced. Ammonia produced in this way is widely used in industry to make nitric acid, fertilizers and drugs. Ammonia is also used by the electronics industry as a blanketing medium for preparing transistors.

NITROGEN CYCLE

Plants make their proteins from nitrates in the soil. These nitrates are partly replenished by the following cycle of events, known as the "nitrogen cycle."

Plant proteins are eaten by animals, including humans, and converted into the particular proteins which the animal requires. Some animals get their proteins indirectly from plants by eating other animals. The waste products of the animals, if returned to the soil, are broken down into ammonium salts which are converted by bacteria in the soil first into nitrites and then into nitrates. This cycle is shown at the top of page 135.

Of course, much sewage is not returned to the soil and so, to maintain the cycle, fertilizers containing ammonium salts and nitrates must be used. Some nitrate is replaced naturally after thunderstorms. Lightning causes a little nitrogen and oxygen in the air to combine to form nitric oxide.

$$N_2 + O_2 \rightleftharpoons 2NO$$

When this oxide cools it combines with oxygen in the air to form nitrogen dioxide.

$$2NO + O_2 = 2NO_2$$

This dissolves in rainwater to form nitric acid.

$$3NO_2 + H_2O = 2HNO_3 + NO$$

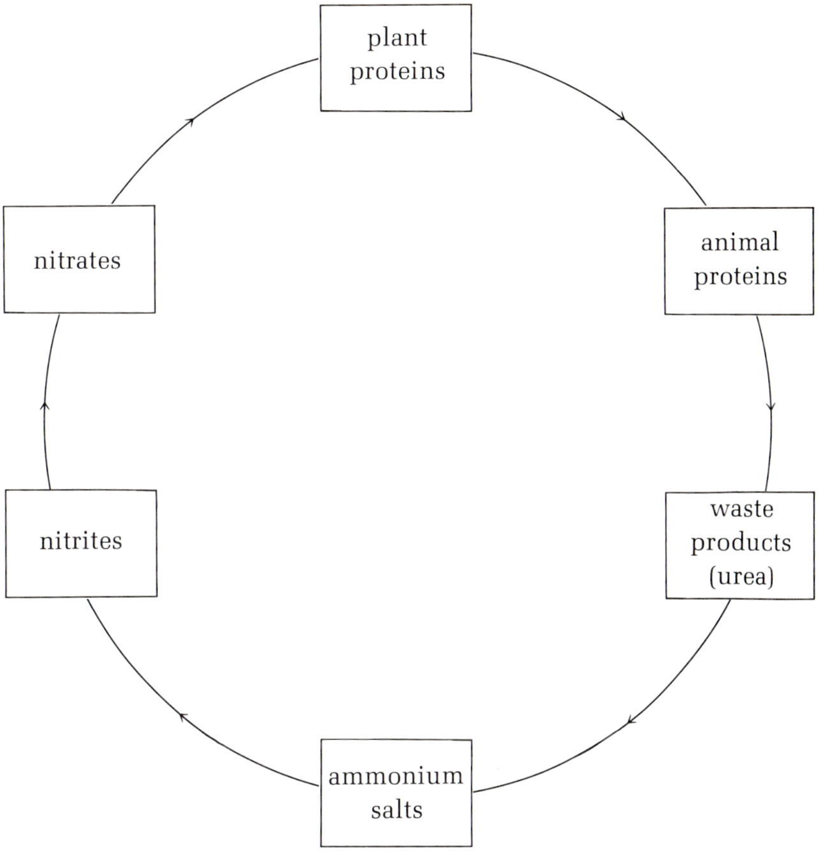

This acid then reacts with substances in the soil, for example, carbonates to produce nitrates.

$$CaCO_3 + 2HNO_3 = Ca(NO_3)_2 + H_2O + CO_2$$

Most plants can only use nitrogen if it is provided for them in the form of inorganic salts, namely, nitrates (salts of nitric acid, HNO_3), nitrites (salts of nitrous acid, HNO_2) and ammonium compounds. There are, however, certain plants which are able to use nitrogen from the air. These are the leguminosae, which include peas, beans, vetches. On the roots of these plants little nodules can be seen (Fig. 18.1). These contain nitrogen-fixing bacteria. They convert the elemental nitrogen from the air into inorganic nitrogen-containing salts which can then be used by the plants.

Soil bacteria thus play a vital part in the nitrogen cycle, both by fixing nitrogen from the air and also by breaking down dead animal and

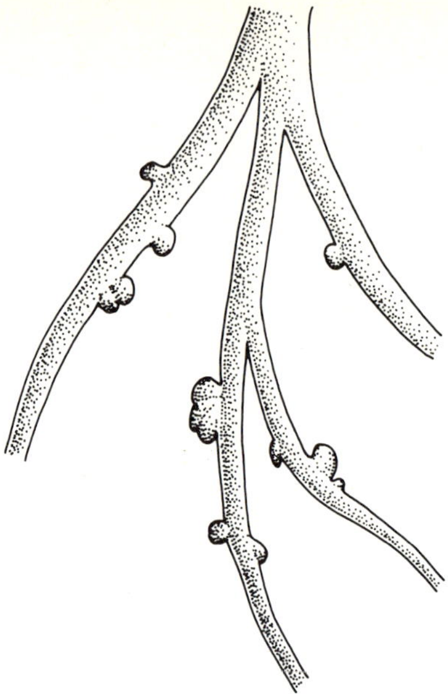

Figure 18.1
Root nodules.

vegetable matter. Without bacteria life on this earth would not be possible.

NITROGEN IN THE BODY

See Chapter 24 on proteins.

PHOSPHORUS

Phosphorus (Gr. *phosphoros,* light-bearing) atoms, can be arranged in several different ways so that there are a number of different physical forms, or *allotropes* (Gr. *allo,* other + Gr. *tropos,* form), of the element: the three principal ones being white (or yellow), red (or violet), and black. The element never occurs free in nature but its compounds are widely distributed. It occurs in minerals: for example phosphate rock is an important source of the element. It also occurs in all living organisms, both animal and vegetable. It is an essential ingredient of cell protoplasm, blood plasma, bone, and brain.

Phosphorus was discovered by Brand in 1669 who prepared it from urine. White phosphorus is a waxy solid, usually yellowish white in color but transparent and colorless when pure. It is highly poisonous, about 50 mg being a fatal dose. If exposed to air it catches fire spontaneously. It must therefore be kept under water and handled with forceps in order to avoid burns. White phosphorus is phosphorescent. If exposed to sunlight or when heated in its own vapor it is converted to red phosphorus. In this form it is no longer phosphorescent, nor does it catch fire spontaneously and it is less dangerous. Red phosphorus is used in the manufacture of safety matches, fireworks, incendiary shells, smoke bombs, tracer bullets, pesticides, and phosphoric acid, H_3PO_4. Large amounts of phosphoric acid are used to make superphosphate fertilizers. Bone ash, which is mainly calcium phosphate, $Ca_3(PO_4)_2$ is used in the production of bone china.

Mono-, di- and tri-phosphates

These are compounds with the following general formulas:

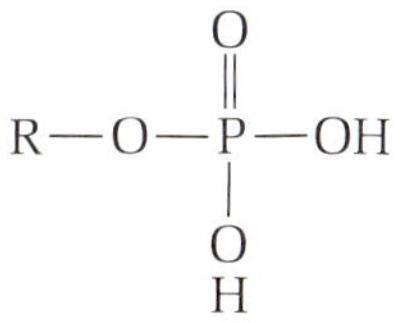

MONOPHOSPHATE
REFERRED TO LATER AS R—Ⓟ

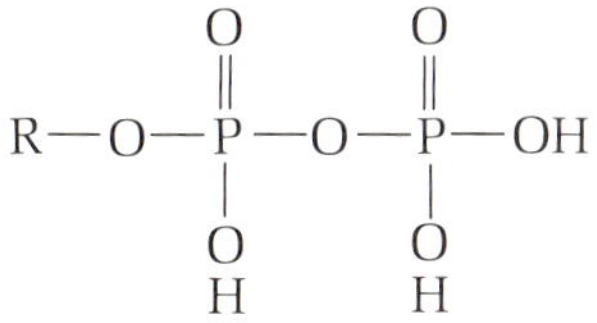

DIPHOSPHATE
REFERRED TO LATER AS R—Ⓟ—Ⓟ

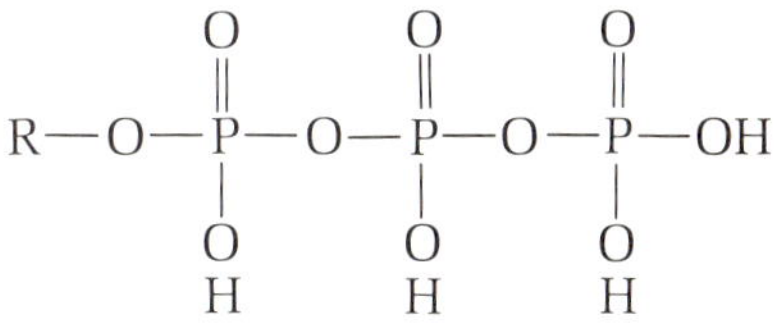

TRIPHOSPHATE
REFERRED TO LATER AS R—Ⓟ—Ⓟ—Ⓟ

R represents a group of atoms. Important examples in body chemistry are AMP (adenosine monophosphate), ADP (adenosine diphosphate) and ATP (adenosine triphosphate).

PHOSPHORUS IN THE BODY

The minimum phosphorus requirement for an adult is approximately 1 g (32 mmol) daily. The main dietary source is provided by phosphates, both inorganic and organic. Foods which are rich in phosphorus include eggs, milk, cheese, almonds, and fish, especially sardines. The phosphorus in flour is mainly unavailable due to the presence of phytic acid which interferes with the absorption of phosphates, calcium and iron. Inorganic phosphate is readily absorbed from the gut in regions where the reaction is slightly acid. An alkaline medium tends to convert phosphates into an insoluble form such as calcium phosphate.

The normal serum inorganic phosphate is 0.8 to 1.4 mmol/liter (2.5 to 4.4 mgP/100 cm^3). Organic phosphorus compounds are more abundant in the red cells than in the serum. Phosphorus is excreted mainly by the kidneys. About $\frac{2}{3}$ of the daily intake is excreted in the urine and so the amount excreted varies considerably, according to the diet, namely 11 to 32 mmol/24 hr. (340 to 1000 mgP/24 hr.).

Phosphates are intimately involved in the calcification of bones and teeth whose rigidity is essentially due to impregnation of the organic collagen matrix by hydroxy-apatite, $Ca(OH)_2 \cdot 3Ca_3(PO_4)_2$. The enzyme phosphatase plays an important part in this calcification process. Parathormone, the hormone secreted by the parathyroid glands, mobilizes calcium and phosphate ions from the bones, and promotes the excretion of phosphate in the urine. Excess secretion of parathormone, as in a parathyroid tumor, causes excess mobilization of mineral from the bones, resulting in osteitis fibrosa cystica with weakening of the bones.

Adequate vitamin D in the diet is required to promote the absorption of calcium from the gut together with the maintenance in the blood of normal calcium and phosphate levels. In rickets due to vitamin D deficiency in the diet and osteomalacia due to poor intestinal absorption of vitamin D, the serum calcium and phosphate are both low. The bones are unduly soft due to inadequate hydroxy-apatite deposition and so skeletal deformities result.

The pH of the blood is maintained within very narrow limits. Phosphate buffers provide an important mechanism for maintaining this constancy of the pH. Plasma contains dihydrogen phosphate ions, $H_2PO_4^{\ominus}$ and the less acid monohydrogen phosphate ions, $HPO_4^{2\ominus}$ in the ratio of about 1:5. In the urine the ratio is reversed and is normally 9:1, but when there is excess acid to be eliminated the ratio may rise to 50:1. This appears to be achieved by the following reaction occurring in the proximal tubules of the kidney:

$$\underset{\text{MONOHYDROGEN PHOSPHATE ION}}{HPO_4^{2\ominus}} + \underset{\text{CARBONIC ACID}}{H_2CO_3} \longrightarrow \underset{\text{DIHYDROGEN PHOSPHATE ION}}{H_2PO_4^{\ominus}} + \underset{\text{BICARBONATE ION}}{HCO_3^{\ominus}}$$

The dihydrogen phosphate ions are excreted in the urine and the bicarbonate ions are returned to the plasma to provide additional base reserve.

Organic phosphorus compounds are numerous and include phosphoproteins (especially in the brain), phospholipids (in every cell in the body), and the adenosine phosphates which provide the energy transfer currency in cellular respiration.

19

Iron and Related Metals

There are a number of metals in the body that are of great importance although they are present in relatively small quantities. Of these iron is present in the greatest amount, enough to make a fairly large nail. It is best known as an essential part of the hemoglobin molecule, required for the transport of oxygen, but it is also involved in a number of enzyme systems, particularly those involved in tissue respiration. Other vital metals include manganese, cobalt, copper, and zinc. The importance of some of these metals has only been appreciated fairly recently. Cobalt, for example, is an essential element in vitamin B_{12} (cyanocobalamin). Inadequate absorption of vitamin B_{12} leads to pernicious anemia.

The metals named above are chemically related. They all belong to the same group of 10 elements forming the first "transition series" of the periodic table.

ATOMIC STRUCTURE

Manganese, iron, cobalt, nickel, and zinc all have two electrons in their outer (4th or N) electron shell. The difference between them in electron configuration lies in the number of electrons in their 3rd or M electron shells. Copper belongs to the same family, although it has only one electron in its outer shell. The gradual transition in their atomic structure is clearly shown below

ELEMENT	ATOMIC NUMBER	NUMBER OF ELECTRONS IN SHELL K (1st)	L (2nd)	M (3rd)	N (4th)
manganese	25	2	8	13	2
iron	26	2	8	14	2
cobalt	27	2	8	15	2
nickel	28	2	8	16	2
copper	29	2	8	18	1
zinc	30	2	8	18	2

Their similarity in atomic structure is reflected in the similarities in their physical and chemical characteristics.

PROPERTIES

Hardness

These metals are all rather hard metals and being relatively stable, they provide useful structural material, either as elemental metals or as alloys, for example, bronze (copper, tin), brass (copper, zinc), and stainless steel (iron, carbon, chromium and sometimes nickel and cobalt).

Valence

As might be expected from the number of electrons in the outer N shell, copper can have a valence of 1 and the rest a valence of 2. But apart from zinc all these elements exhibit more than one valence. This is because some electrons in the 3rd or M shell can also act as valence electrons. With zinc this does not happen, so it has a valence of 2 only.

Color of compounds

These metals can form colored complex ions, for example, potassium hexacyanoferrate(III) (potassium ferricyanide) is ruby red and owes its color to the complex hexacyanoferrate(III) (ferricyanide) ion, $Fe(CN)_6^{3\ominus}$ which has a three-dimensional structure like this:

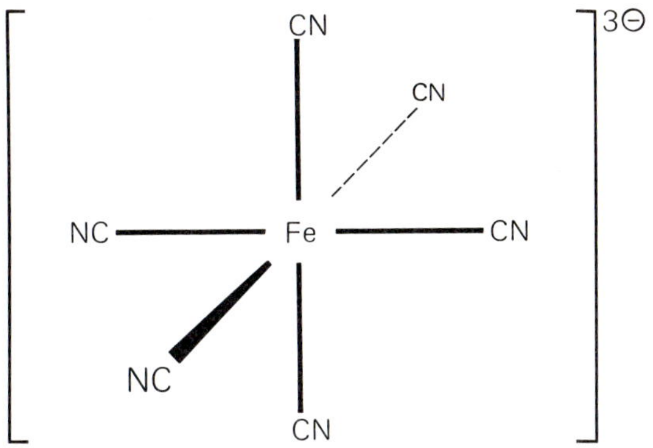

The color is due to the ability of these complex ions to absorb light of certain wavelengths in the visible spectrum. The remaining light is deprived of the absorbed wavelengths and gives the substance its color. To be able to absorb light radiation a substance must have an electronic structure which is able to accept the additional energy involved. With the exception of zinc the ions have just such a structure. Electrons are raised to a higher energy level by the absorbed radiation. If no radiation is absorbed a substance appears white under normal lighting.

Catalytic activity

These metals can act as catalysts, either in elemental form, for example iron accelerating the formation of ammonia from nitrogen and hydrogen, or in simple inorganic compounds, for example manganese dioxide, MnO_2 facilitating the release of oxygen from heated potassium chlorate, or else in complex organic molecules such as cytochrome oxidase, an enzyme containing iron which facilitates the utilization of oxygen in tissue respiration. In the body's chemistry these complex catalysts, known as enzymes, play a very important part.

MANGANESE

Manganese (L. *magnes*, magnet) is a gray-white metal resembling iron but harder and very brittle. It is used in alloys, for example, in steel, to which it gives great toughness, and in ferromagnetic alloys. Amethysts owe their purple-violet color to the presence of manganese.

In the body, manganese is present in a considerable number of enzymes, for example, choline esterase, the enzyme responsible for removing acetyl choline as soon as soon as it has completed one of its many functions such as the conduction of stimuli from nerves to muscles, causing the muscles to contract. Without choline esterase the muscles would be unable to relax again.

Manganese deficiency occurs in animals, causing sterility in male rats and failure of lactation in female rats; pregnant sows with manganese deficiency may reabsorb the fetus or produce undersized weakly offspring; in chicks manganese deprivation causes deformities of the legs and spine. Manganese deficiency has not been demonstrated in man. Excess manganese interferes with iron absorption, causing iron deficiency anemia.

Potassium permanganate, $KMnO_4$

Potassium permanganate is a powerful oxidizing agent and as such is an effective antiseptic. Crystals of potassium permanganate are purple prisms which dissolve in water to give a purple solution. When used as an antiseptic foot bath it tends to stain the feet a brownish color due to the formation of brown manganese dioxide, MnO_2. The oxidizing action of potassium permanganate is related to the readiness with which the permanganate ion $MnO_4^{\ominus}$ accepts electrons in the presence of hydrogen ions or water molecules. Its USP category is topical anti-infective.

IRON, Fe

Iron (Anglo-saxon, *iron*), plays an important part in our body's chemistry and in our daily lives. It is widely distributed throughout the universe, occurring in many types of stars, including our sun. In the earth's crust, it is the fourth most abundant element by weight and is probably one of the main constituents of the earth's core.

Extraction of iron from its ores dates back to the early Eastern Mediterranean civilizations of about 2300 B.C. The extraction process, known as smelting, consists essentially of the reduction of iron oxides, such as magnetite, Fe_3O_4 and hematite, Fe_2O_3, or iron(II) carbonate (ferrous carbonate) such as siderite, $FeCO_3$ to metallic iron in varying degrees of purity. The reduction is accomplished by carbon monoxide that is produced by heating coke in a limited air supply.

$$\underset{\text{FERRIC OXIDE}}{Fe_2O_3} + \underset{\text{CARBON MONOXIDE}}{3CO} \longrightarrow \underset{\text{IRON}}{2Fe} + \underset{\text{CARBON DIOXIDE}}{3CO_2}$$

In compounds, iron (L. ferrum) can have a valence of either two or three. The term ferr*ous* indicates a valence of two, for example, ferrous chloride, $FeCl_2$, and ferr*ic* indicates a valence of 3, for example, ferric chloride, $FeCl_3$. Nowadays ferrous and ferric compounds are called iron(II) and iron(III) compounds, respectively, the roman numerals indicating the valence. Ferrous, iron(II), salts are much more readily absorbed from the intestine than are ferric, iron(III) salts. So oral preparations for the treatment of iron deficiency anemia are almost always iron(II) compounds, the one most commonly used being iron(II) sulfate.

Iron(II) sulfate

The crystals of iron(II) sulfate contain 7 moles of water of crystallization to each mole of the salt ($FeSO_4 \cdot 7H_2O$). Crude iron(II) sulfate is known as Green Vitriol or Green Copperas. In purified form it consists of transparent green crystals. When dissolved in water it forms an acidic solution, pH less than 7.

Tablets of iron(II) sulfate (e.g. Fersolate, Glaxo) provide one of the most convenient and economical ways of prescribing iron for adults and are usually well tolerated. Its USP category is "hematinic." Constipation is occasionally produced. Larger doses can cause diarrhea. By adjustment of the dose an acceptable balance can usually be obtained.

Gastro-intestinal disturbances can be minimized by film-coated tablets which only release the iron(II) sulfate slowly, as in Ferrogradumet (Abbott Labs.). Ferrogradumet C (Abbott Labs) is similar but contains vitamin C to aid iron absorption.

On occasions sugar-coated tablets of iron(II) sulfate have been taken by children as sweets, with disastrous results. Large doses have frequently proved fatal. Smaller doses can cause severe gastrointestinal irritation with hematemesis (vomiting of blood), shock and rapid pulse. The toxicity of iron(II) sulfate is largely due to the fact that it is ionized into $Fe^{2\oplus}$ and $SO_4^{2\ominus}$ and being soluble gives a high concentration of iron(II) ions. The effect on the body is most marked in children, for whom alternatives such as iron(II) carbonate, iron(II) gluconate or Sytron (a complex iron(III) salt, Parke Davis) are usually given.

Iron(II) carbonate, $FeCO_3$

Iron(II) carbonate is insoluble in water but dissolves slowly in the stomach and gives a low concentration of iron(II) ions. Ferrodic (Allen and Hanbury) tablets contain iron(II) carbonate and vitamin C. These tablets were used in a trial involving 160 children conducted by one of the authors and found to be a satisfactory way of administering iron.

Iron(II) gluconate

Iron(II) gluconate ionizes only slightly in solution. If the gluconate radical is represented by Gl, the formula of iron(II) gluconate may be expressed as $FeGl_2 \cdot 2H_2O$.

The full formula is

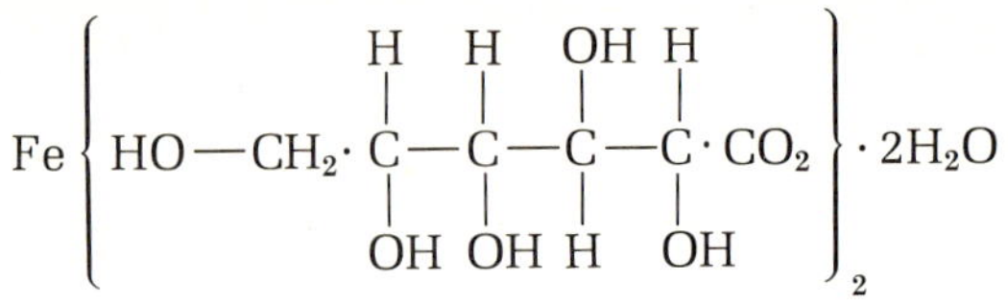

FERROUS (IRON(II)) GLUCONATE

Parenteral iron

Where absorption of iron from the intestine is seriously impaired, it is necessary to have iron preparations which can be injected. Imferon (Lakeside Labs) is one example. It consists of iron(III) hydroxide, $Fe(OH)_3$ combined with dextrans. It was originally designed for intramuscular injection but was found to cause brown staining at the site of injection. So it is now usually given intravenously, often in the form of an infusion with normal saline (5% dextrose has been used instead of saline but can lead to venous thrombosis). Animal experiments have indicated that iron injections might cause tumor formation but this has not been observed in humans. Jectofer (Astra Pharmaceutical Prod., Inc.) is another iron preparation which is widely used for injecting or infusion. Its chemical composition is complicated (iron sorbital citric acid complex).

Iron metabolism

Iron is present in meat, especially liver and kidney, also eggs, cocoa, chocolate, prunes, and some vegetables, such as spinach. In general, the absorption of iron from organic sources tends to be poor but there are exceptions such as beef. Iron from white flour is more readily available than from whole-meal flour. Inorganic phosphate in the diet impairs the absorption of both calcium and iron by promoting the formation of insoluble salts.

It should be noted that milk contains extremely little iron. In some countries infants are breast-fed for 2 to 3 years and develop iron-deficiency anemia. The newborn infant has more hemoglobin than it requires. So the excess red cells are broken down and the iron is stored. This stored iron normally tides the infant over the first few months of life until the child is weaned.

During growth, iron is required for the formation of hemoglobin in the red cells, myoglobin in the muscles, cytochrome in all cells and a number of other enzymes. In males, the iron requirement is greatest at 15 to 16 years. In females, the iron requirement is greatest during the course of reproductive life to replace iron lost during menstruation and pregnancy. It is during the last three months of pregnancy that the fetus takes the greatest amount of iron from the mother, doubled in the case of twins. For an adult male 90 μmol (5 mg) of iron a day is more than adequate but this needs to be doubled for females, the exact amount depending on the size of the menstrual loss and the requirement being reduced after the menopause.

Absorption of iron is mainly from the small intestine. Studies with radioactive iron ^{59}Fe, have shown that absorption of iron is more than normally efficient in people with iron deficiency. This is because the normal person is protected against absorbing too much iron by a mechanism which is essentially chemical. Iron is absorbed into the cells of the intestinal mucosa in the iron(II) ($Fe^{2\oplus}$) state. When it has been absorbed, it is oxidized into the iron(III) ($Fe^{3\oplus}$) state in the form of iron(III) hydroxide $Fe(OH)_3$. This combines with a protein, apoferritin, to form an iron-protein complex, ferritin. In this form it is temporarily stored in the mucosal cells until needed (Fig. 19.1). If no free apoferritin is available, then the absorption of iron stops. Iron is unusual in that the amount excreted is very small and cannot be regulated. So the only method of controlling the body iron is at the point of absorption.

Iron is stored in the liver, also in the form of ferritin. The transfer of iron from intestinal mucosa to the liver and also the tissue cells involves a plasma protein (transferrin). The tissue which takes up most of the iron is the bone marrow where it is required for the formation of red cells. It

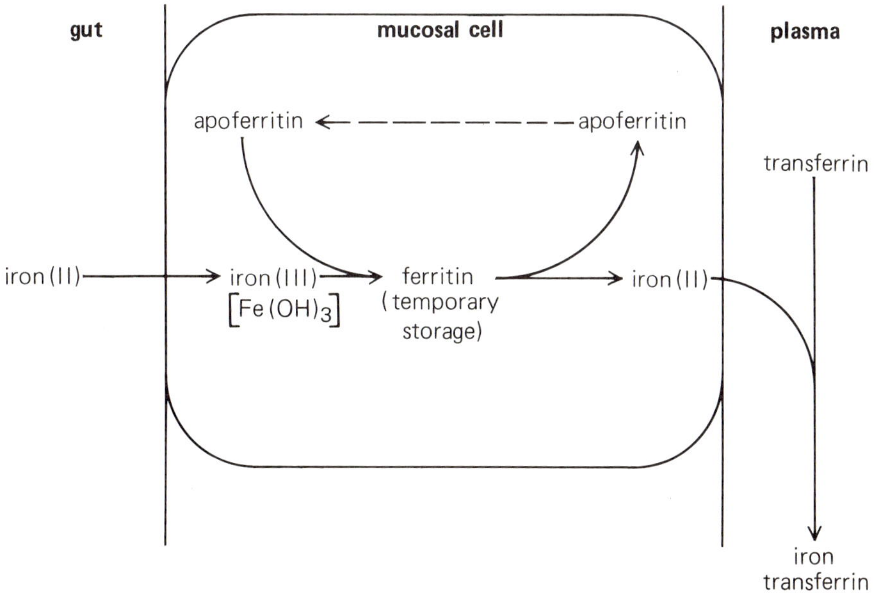

Figure 19.1
Iron absorption by mucosal cell of gut.

should be noted that the main source of iron for making new red cells is from the breakdown of old cells. The iron is continually being recycled. Only a very small proportion of the total iron turnover is usually from the diet, just enough to make good any loss, for example, from hemorrhage, provided this is not excessive. Any severe hemorrhage soon leads to iron deficiency.

The total iron content of an average man is about 80 mmol (4.5 g). Of this about 60% is in red cells (hemoglobin), 20% is in muscles (myoglobin) and tissue enzymes (cytochrome oxidase, peroxidase, catalase) and 20% is stored in the liver, spleen, and bone marrow (ferritin and hemosiderin).

In untreated, iron-deficiency anemia the serum iron is reduced, the normal being 1 to 3 mmol/liter (60 to 180 $\mu g/cm^3$). The level is raised in hemochromatosis and infective hepatitis. The iron in serum is bound to the protein transferrin. The amount of the latter determines the iron-binding capacity of the serum which is normally 4.5 to 8 mmol/liter (250 to 450 $\mu g/cm^3$.) The iron-binding capacity is increased in iron deficiency and is one of the more reliable indications that this condition is present. The amount of iron present, expressed as a proportion of the iron-binding capacity, is the iron saturation. Normally, it is 15 to 40%. In iron deficiency it is usually less than 10%. In uncomplicated pernicious anemia and hemochromatosis, it is increased to nearly 100%.

Hemoglobin

Hemoglobin has a complex molecular structure incorporating iron. It can bind molecular oxygen in a loose and easily reversible combination. The iron itself is not oxidized and remains in the iron(II) state throughout. The symbol Hb represents unoxygenated (reduced) hemoglobin and HbO_2 represents oxyhemoglobin. The chemical structure of hemoglobin is described in Chapter 25.

In iron deficiency anemia there is insufficient iron available for the bone marrow to synthesize adequate hemoglobin. So the red cells are smaller than normal (microcytic) and incompletely filled with hemoglobin (hypochromic); see Fig. 19.2.

Carboxyhemoglobin

Carbon monoxide has an even greater affinity for hemoglobin than oxygen, combining with it firmly. The carboxyhemoglobin which is formed blocks the hemoglobin molecule for oxygen carriage, thus depriving the tissues of oxygen. Relatively small amounts of carbon monoxide in the atmosphere, as from a car exhaust in a closed garage, can cause serious or fatal effects.

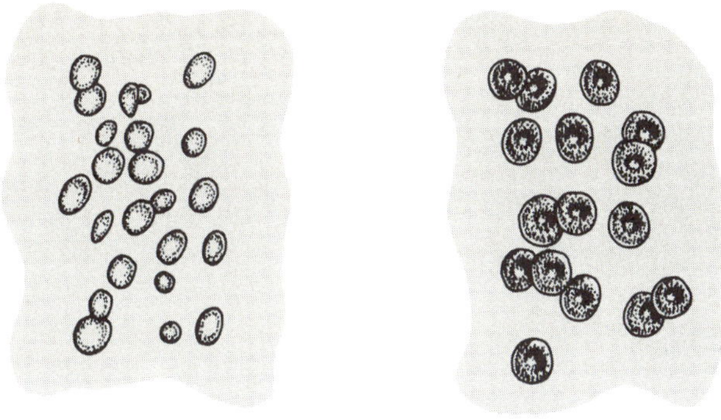

(a) (b)

Figure 19.2
(a) Red cells in iron deficiency tend to be small with less hemoglobin than normal; (b) normal red cells.

COBALT

Cobalt (Ger. *Kobold*, goblin) is a hard, brittle metal resembling iron and nickel in appearance. It is used in making stainless steel and specially tough alloys, for example, for jet engines. One application is in the manufacture of surgical instruments, enabling them to be sterilized in a flame without damage. Like iron it has two valences. Its valence is two in cobalt*ous* (cobalt(II) compounds, which includes all its simple compounds, and three in cobalt*ic* (cobalt(III)) compounds which are more complex. Salts of cobalt have been used for centuries to provide beautiful and permanent blue colors in glass, porcelain, pottery, tiles, and enamel. If a solution of cobalt(II) chloride, $CoCl_2 \cdot 6H_2O$ is used as ink, the very faint pink is invisible. When the paper is warmed a clearly visible blue color develops because the water of crystallization is driven off, leaving a blue anhydrous salt. It soon rehydrates from the water vapor in the air and becomes invisible again.

Cobalt deficiency causes "pine" in sheep and cattle in whom vitamin B_{12} is normally formed by bacteria during rumination. All nonruminating animals, including man, require vitamin B_{12} in their diet instead of cobalt. We depend on animals and plants for our vitamin B_{12} and all eventually depend on soil bacteria to convert simple cobalt compounds into vitamin B_{12}.

Vitamin B_{12}

Vitamin B_{12} (cyanocobalamine) is a very complex molecule with cobalt at its center (Fig. 19.3). Liver is a rich source. In 1926, Minot and Murphy showed that patients with pernicious anemia, who would have died, improved on large quantities of raw liver by mouth. Later, extracts

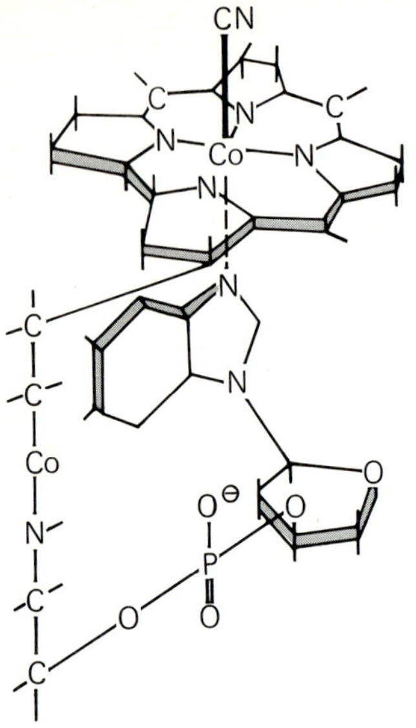

Figure 19.3
Skeleton of vitamin B_{12} (cyanocobalamine) molecule.

from liver were prepared which could be injected. Crystalline vitamin B_{12} was finally isolated in 1948, independently in Britain and America, and was shown to be the effective agent in the treatment of pernicious anemia.

The essential cause of pernicious anemia is failure of the stomach to produce a substance (intrinsic factor) necessary for the absorption of vitamin B_{12} in the small intestine. After absorption, vitamin B_{12} is stored in the liver and utilized by the tissues, especially the bone marrow.

Vitamin B_{12} is an essential enzyme in the complicated nucleotide syntheses involved in the formation of new cells. In the bone marrow new cells must continually be produced to replace the time-expired cells, otherwise a progressive, pernicious anemia develops with a deficiency not only of red cells (Fig. 19.4) but also of white cells and platelets. In addition, vitamin B_{12} is required by the nervous system and without it, degenerative changes may occur (subacute combined degeneration of the cord) with progressive paralysis, loss of sensation, mental deterioration and even dementia.

COPPER

Metallic copper (L. *cuprum*, Cyprus, an island rich in copper) has a characteristic color to which it gives its name, with a bright metallic luster when polished. It is malleable, and ductile and a good conductor of heat and electricity. Its alloys include bronze, brass, gunmetal, and coin metals. Copper has two valences. In cuprous (copper(I)) compounds, it has a valence of one, as in copper(I) chloride, CuCl, and in

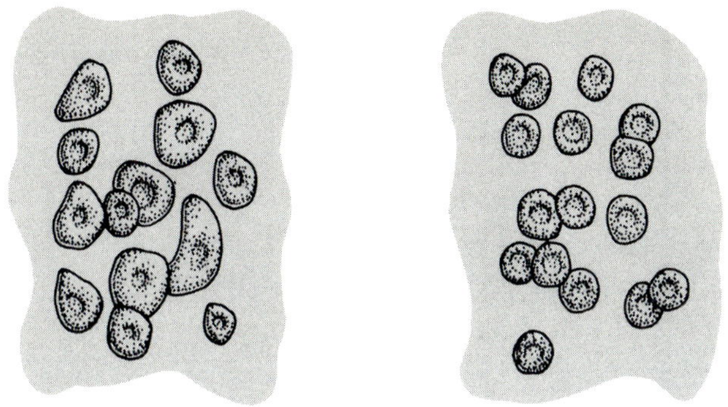

(a) (b)

Figure 19.4
(a) Red cells in pernicious anemia tend to be larger and more variable in size and shape than normal; (b) normal red cells.

cupric(copper(II)) compounds, a valence of two, as in copper(II) chloride, $CuCl_2$. Copper(II) chloride may be made by passing excess chlorine over heated copper.

$$\underset{(s)}{Cu} + \underset{(g)}{Cl_2} = \underset{(s)}{CuCl_2}$$

Strong heating causes copper(II) chloride to break down into copper(I) chloride and chlorine.

$$\underset{(s)}{2CuCl_2} = \underset{(s)}{2CuCl} + \underset{(g)}{Cl_2}$$

Of the compounds of copper, the one most widely used is copper(II) sulfate, commonly called copper sulfate.

Copper(II) sulfate, $CuSO_4{\cdot}5H_2O$

Hydrated copper(II) sulfate is a blue crystalline substance, known as blue vitriol. It contains 5 moles of water of crystallization for each mole of salt ($CuSO_4{\cdot}5H_2O$). It is prepared commercially by spraying dilute sulfuric acid on scrap copper in a tower up which air is blown.

$$\underset{\substack{(s)\\ \text{COPPER}\\ \text{SCRAP}}}{2Cu} + \underset{\substack{(aq)\\ \text{SULFURIC}\\ \text{ACID}\\ \text{(DILUTE)}}}{2H_2SO_4} + \underset{\substack{(g)\\ \text{OXYGEN}\\ \text{(FROM AIR)}}}{O_2} = \underset{\substack{(aq)\\ \text{COPPER(II)}\\ \text{SULFATE}}}{2CuSO_4} + \underset{(l)}{2H_2O}$$

Copper(II) sulfate dissolves readily in water to give a blue solution. If this is evaporated the blue crystals of copper(II) sulfate pentahydrate ($CuSO_4 \cdot 5H_2O$) are obtained.

Copper(II) sulfate is strongly toxic to the lower organisms but much less so to the higher ones. So in very small quantities it may be added to swimming pools to destroy algae. It is used to prevent the "damping off" of seedlings caused by fungus infection, for spraying vines and for preventing potato blight. It is also an insecticide.

Biological aspects of copper

Copper in the diet assists the absorption of iron. Thus rats on a copper-deficient diet develop a microcytic anemia that responds to injected iron but is resistant to iron by mouth. Infants on a copper-deficient diet also develop a microcytic anemia. The normal level of copper in the serum is 12 to 25 μmol/liter (75 to 160 μg/100 cm^3). This may be reduced in Wilson's disease, a metabolic defect involving copper in which the brain and liver are affected and large amounts of copper are excreted in the urine (over 20 μmol (1.3 mg)/24 hr).

Copper is present in some of the enzymes in the human body, for example tyrosinase. There is a protein containing copper in our red cells (erythrocuprein) but its function is unknown. It is of interest that oxygen carriage in humans is by hemoglobin which contains four atoms of iron in its molecule, but in some creatures, such as squids, it is by a hemocyanin containing up to 20 atoms of copper in its molecule.

ZINC

Zinc (Ger. *zink*, unknown) is a bluish-white lustrous metal, brittle at ordinary temperatures but malleable at 100°C to 150°C. It is used in many alloys, for example, brass, German silver, and soft solder.

Zinc is resistant to atmospheric corrosion because it forms a protective film of carbonate on its surface, from the carbon dioxide and oxygen in the air. So iron and steel may be protected from corrosion by dipping the cleaned metal into molten zinc (galvanizing) so that a thin covering of zinc is obtained.

Unlike the rest of the transition metals, zinc has a fixed valence of two. This is determined solely by the two electrons in its N shell. The M shell is stable, being completely filled by 18 electrons. Zinc sulphide is used for X ray and T.V. screens, fluorescent lights and luminous dials. Zinc oxide, however, is the compound with the widest applications.

Zinc oxide, ZnO

Zinc oxide is an insoluble white powder. It is probably most familiar in the form of zinc oxide ointment, used for skin application as a soothing cream. Its USP category is "astringent and topical protectant." Zinc oxide is used in the manufacture of pharmaceuticals, cosmetics, paints, rubber products, floor coverings, plastics, printing ink, soap, textiles, and electrical equipment. It has unusual physical (optical, thermal, electrical, and solid state) properties which are not yet fully understood.

Zinc sulfate

Zinc sulfate consists of odorless, colorless crystals that turn to fine powder on exposure to the air. They have 7 moles of water of crystallization to each mole of salt, $ZnSO_4 \cdot 7H_2O$. It is sometimes used in medicine as an emetic. In capsular form it can be used for the administration of zinc in a case of zinc deficiency. Dosage needs to be carefully controlled because its toxic effect can be severe or fatal.

Red lotion (lotio rubra) contains 1% zinc sulfate and is used as an astringent surface antiseptic, for example, in the treatment of indolent ulcers to promote healing. The USP category of zinc sulfate is "astringent (ophthalmic)."

Zinc carbonate, $ZnCO_3$

Zinc carbonate is a white powder. It is one of the main constituents of calamine, which also contains zinc oxide, iron(III) oxide (giving it a pink color) and phenol (skin anesthetic and antiseptic). Calamine is used as a dusting powder and also as a soothing skin lotion suitable for sunburn.

Zinc in the body

The adult human body contains 2 to 4 g zinc, comparable to the amount of iron present which is 4 to 5 g. In contrast to iron which is mainly (75%) in the blood as hemoglobin, most of the zinc is in the tissue cells, those of the eye having the highest level (0.5%). Zinc is present in a number of enzymes, for example, carbonic anhydrase, but there is still much to be learned about its function and why, for example, there should be a low serum zinc in many cases of lung cancer.

20

Some Carbon Chemistry

In the early nineteenth century chemists thought that substances made by living cells had special properties by virtue of the "life force." The study of substances made by living organisms was called "*organic*" chemistry. A familiar example of an organic chemical is alcohol. Its preparation by the fermentation of grape-sugar has long been known to winemakers. More than a thousand years ago, it was discovered that alcohol could be concentrated by distilling it to make spirits and liqueurs. In 1749, Marrgraf distilled red ants and obtained an acid which he called formic acid (L. *formis*, ant). When an ant stings it injects formic acid, causing pain to humans and capable of killing other insects.

Towards the end of the eighteenth century Scheele isolated from lemons the citric acid which gives them their characteristic flavor. He also isolated malic acid from apples, tartaric acid from grapes, lactic acid from sour milk, and uric acid from urine. In 1773, Roualle isolated urea from human urine. A turning point came in 1846 when Wöhler discovered that urea could be made in the laboratory by evaporating a watery solution of an inorganic substance, ammonium cyanate.

$$NH_4CNO \longrightarrow CO(NH_2)_2$$

AMMONIUM CYANATE — UREA

Wöhler had made an organic chemical without the aid of the life force. He had prepared urea "without requiring a kidney, or an animal, either man or dog," he wrote in a letter to a fellow chemist Berzelius.

At first Berzelius and the other chemists of his time could not believe it. Then in 1850, Kolbe provided further support for the idea by synthesizing acetic acid, the pungent acid of vinegar and sour wine. Gradually it became accepted that the same laws applied to both inorganic and organic chemistry.

ORGANIC CHEMISTRY

One thing that organic substances were found to have in common was that they all contained carbon. So organic chemistry became defined as the chemistry of carbon compounds. Carbon compounds far outnumber the compounds of all the other elements put together. These other compounds are known as *inorganic compounds*.

CARBON, C

Carbon (L. *carbo*, charcoal) is very widely distributed in the sun, stars, comets, planets and some meteorites. Our atmosphere contains 0.3% of carbon dioxide. This does not sound like very much but, in fact, it is the source from which all living organisms, directly or indirectly, obtain the carbon from which the molecules of living matter are constructed. Much of this carbon is returned to the atmosphere in the form of CO_2 as a product of respiration, both by animals and plants. But much of it has been locked up for geological ages in dead vegetable matter, from which coal is derived, and dead animal matter, from which petroleum (hydrocarbons) and chalk (carbonates) are derived. Eventually some of this has been released once more into the atmosphere as CO_2 by man's utilization of coal, petroleum, and limestone. Volcanoes also eject much CO_2 into the atmosphere.

Carbon has an extraordinary capacity for making compounds. The principal raw materials from which carbon compounds are made are coal, oil, wood, and agricultural products. More than a million different carbon compounds have been described, outnumbering the noncarbon compounds of all the other elements put together by about twenty to one. It is this capacity of carbon to combine with other atoms in such an enormous number of different ways which has made life possible.

As well as forming numerous compounds, carbon is also found free in nature. Carbon atoms can be arranged in two different ways and so there are two different physical forms, or allotropes, of the element: diamond, and graphite. Diamond is one of the hardest known materials. Graphite is extremely soft, being used in lubricants and "lead" pencils. The different types of charcoal, such as animal and wood charcoal are composed of tiny graphite crystals.

The atomic structure of carbon may be represented as follows:

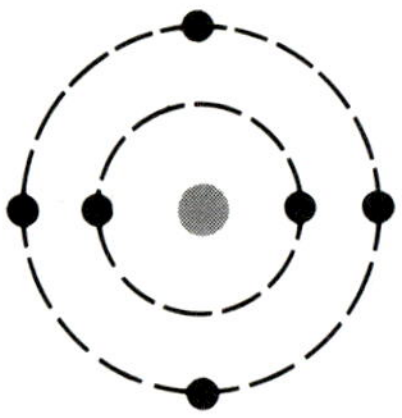

CARBON

Electron configuration	2, 4
Atomic number	6

VARIETY OF CARBON COMPOUNDS

Why does carbon form so many compounds? Carbon has four electrons in its outer shell and so it must make four covalent bonds to acquire a stable octet of electrons. Because of this, and because the bonds made are strong ones, carbon atoms are able to unite with each other and with atoms of other elements (chiefly hydrogen, nitrogen, oxygen, the halogens, sulfur and phosphorus) to form open *chains* of varying length and closed chains, called *rings*, of a variety of shapes and sizes. The number of different ways the atoms can be arranged in these chains and rings is practically infinite. To illustrate the variety of atomic arrangements in carbon compounds, we will consider examples.

STRUCTURE OF CARBON COMPOUNDS

Short chains

Short chains are one type of arrangement as in butane (used as a camping gas) whose structure may be written as

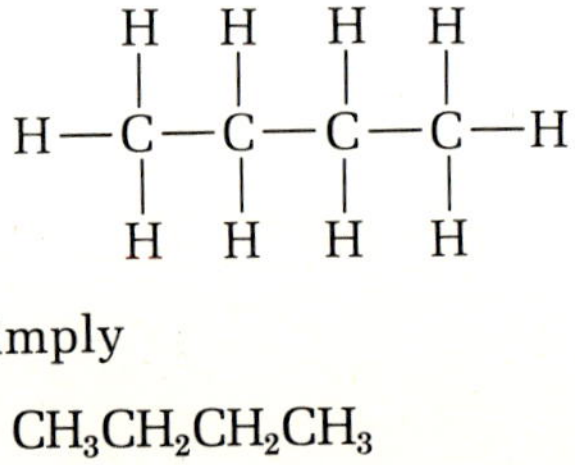

or simply

$CH_3CH_2CH_2CH_3$

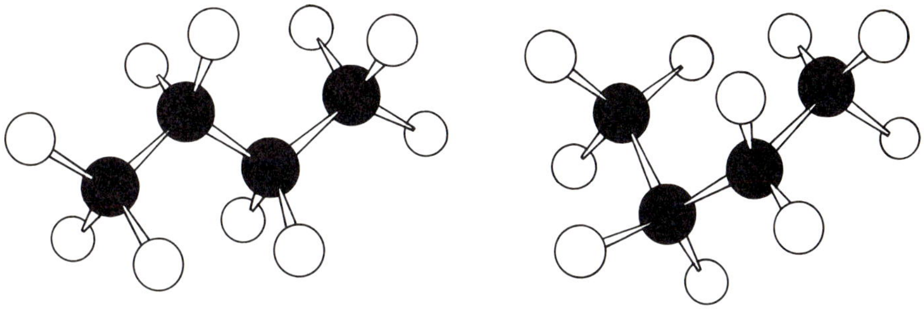

Figure 20.1
Conformations of butane. The carbon atoms are represented by the black spheres and the hydrogen by the white.

These structural formulas do not represent the shape of the molecule, which is in fact continually changing. Some of the possible shapes or *conformations* are illustrated by the models in Fig. 20.1. Notice that the carbon chain is not straight as implied by the formulas. We will return to this important point later when considering the shape of the methane molecule.

Butane is a member of a series of hydrocarbons (compounds containing carbon and hydrogen only) known as *paraffins* or *alkanes* which have the general formula C_nH_{2n+2}. Putting $n = 1$ gives the formula of methane CH_4, the simplest of all organic molecules and the chief constituent of natural gas. The second member of this series ($n = 2$) is ethane C_2H_6 which has the structure

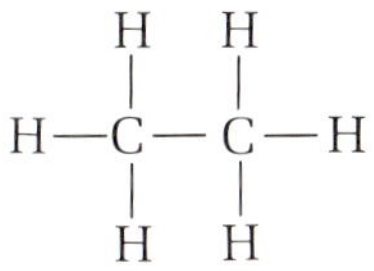

or

CH_3CH_3

Branched chains

Branched chains are a slightly more complicated arrangement as in isobutane.

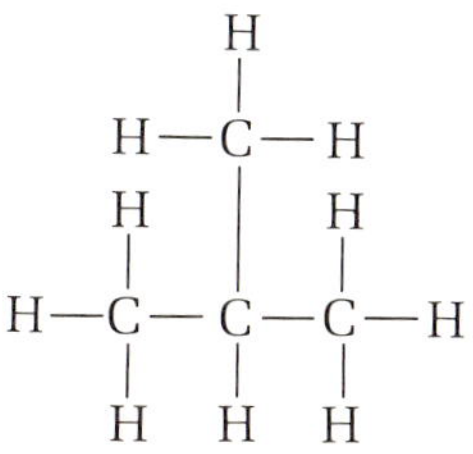

or

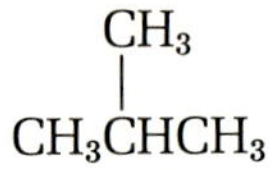

We notice that this alkane has the same molecular formula as butane, namely, C_4H_{10}. The existence of compounds with the same molecular formula is known as *isomerism* and such compounds are called *isomers* (Gr. *isos*, equal + Gr. *meros*, part). See if you can work out the structural formulas of isomers which have the molecular formula C_5H_{12}, bearing in mind that each carbon atom must have four bonds and each hydrogen atom one bond.

Medium-length chains

Medium-length chains have a slightly larger molecular structure, as in soap which contains the negative stearate ion

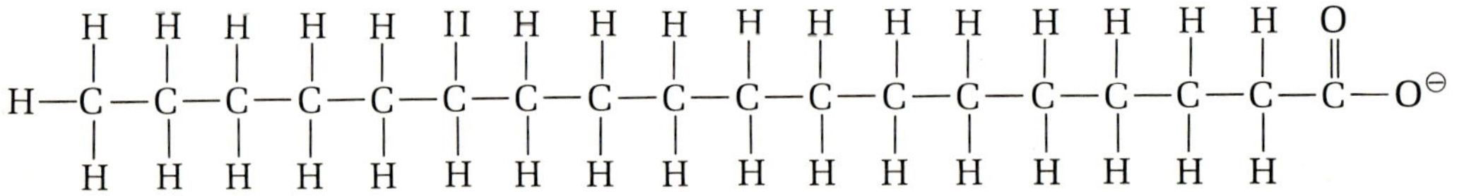

This ion can, of course, be written more simply, if less informatively, as $C_{17}H_{35}COO^{\ominus}$. The grouping $C_{17}H_{35}$ is an example of an alkyl group or radical, which is a group of carbon and hydrogen atoms having the general formula C_nH_{2n+1}. Alkyl groups are found combined with other atoms or groups in the majority of organic compounds.

Putting $n = 1$ in C_nH_{2n+1} gives the formula of the methyl radical CH_3. There are two radicals possible when $n = 3$. One is called normal propyl (or *n*-propyl) and has the structure

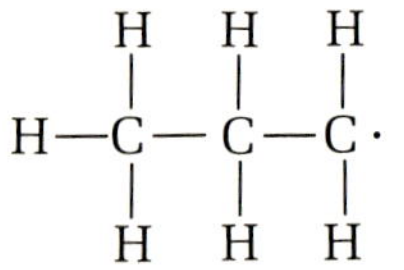

or

$CH_3CH_2\dot{C}H_2$

and its isomer is known as the iso-propyl radical with the structure

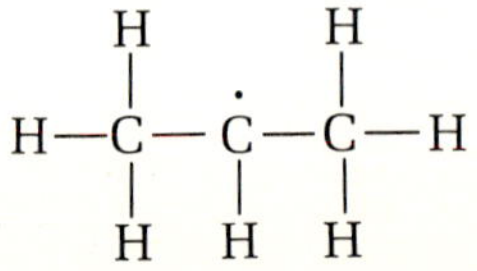

or

$$CH_3\dot{C}HCH_3$$

In these formulas, the dot on a carbon represents an unpaired electron, so that such a carbon atom has only seven electrons in its outer shell and is consequently very reactive. It combines readily with another atom or radical in order to acquire a stable octet. For example, two methyl radicals combine to form ethane.

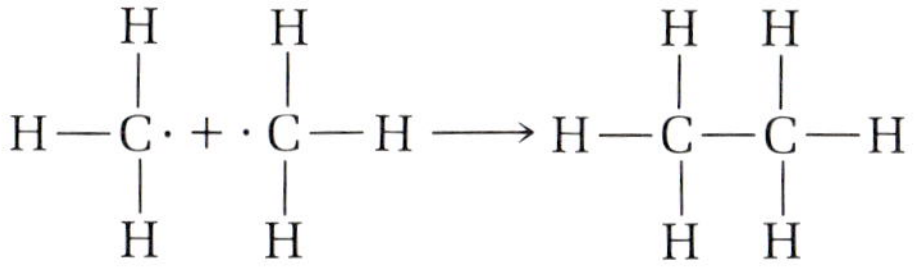

and an ethyl radical combines with a chlorine atom (which we recall has seven electrons in its outer shell) to form ethyl chloride or chloroethane, which is sometimes used as an external local anesthetic, producing its effect by greatly cooling the skin surface as it evaporates.

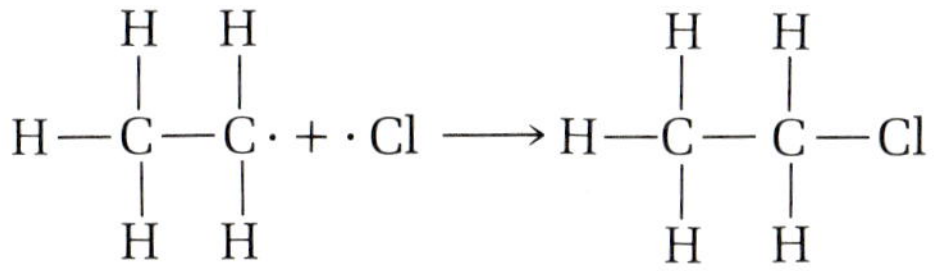

CHLOROETHANE
(ETHYL CHLORIDE)

Radicals, therefore, if they happen to be produced in a chemical reaction, have only a momentary independent existence.

Long chains

Long chains are an arrangement that contains thousands of atoms, as in polyethylene (polythene), which has the following structure:

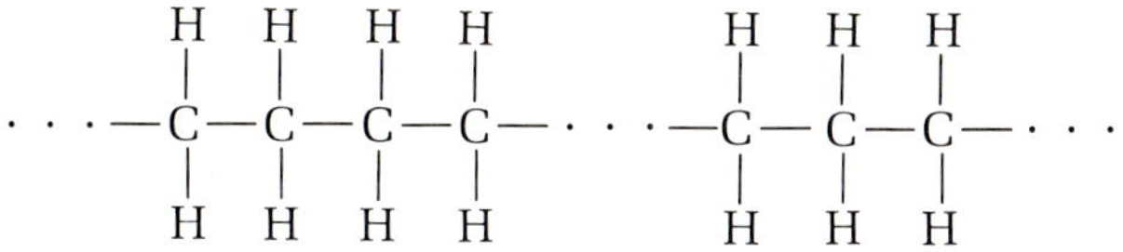

POLYETHYLENE
(POLYTHENE)

This long chain formula is characteristic of plastics.

Rings

Rings show a cyclic arrangement of carbon atoms as in cyclohexane and benzene.

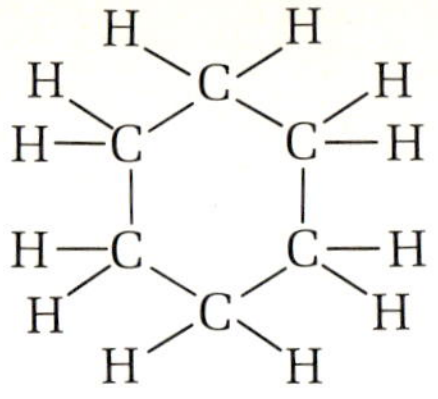

often abbreviated to

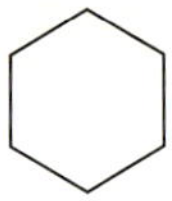

CYCLOHEXANE

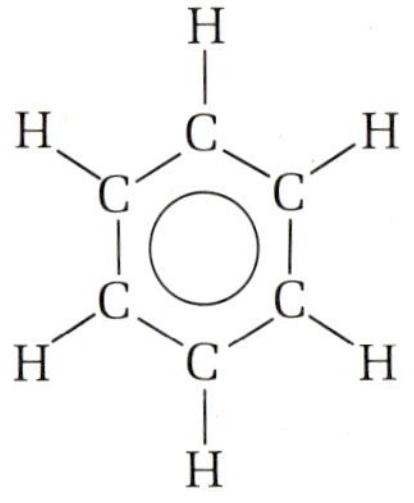

normally abbreviated to

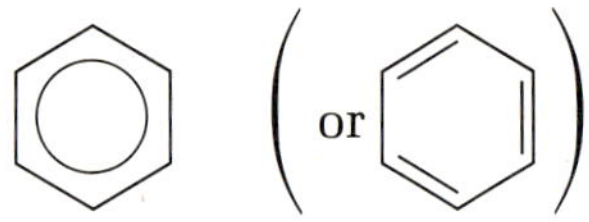

BENZENE

The circle drawn inside the benzene ring of carbon atoms needs an explanation. An uncombined carbon atom has four electrons in its outer shell. In the benzene molecule, three of these electrons are used in forming three covalent bonds, as indicated in the formula. There is much evidence to suggest that the fourth electron on each carbon "wanders" around the ring of six carbon atoms, so that, in all, there are six electrons circulating in the ring. The circle in the structural formula indicates this. These six electrons are said to be *delocalized* because none of them is localized on a particular carbon atom.

Compounds with a similar chemistry to benzene are called *aromatic* and they form a large and very important class of organic compounds. Not surprisingly, the molecules of most of these aromatic compounds contain at least one benzene ring, for example, aspirin and menadione.

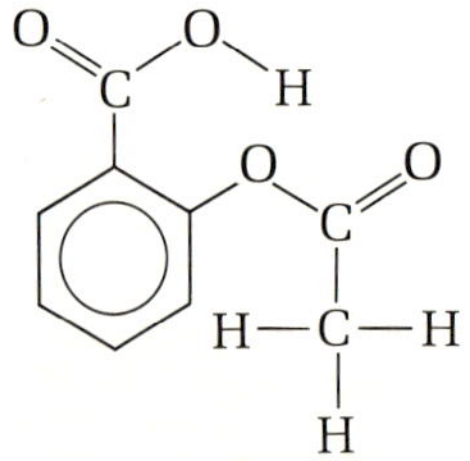

ASPIRIN
(RELIEVES PAIN)

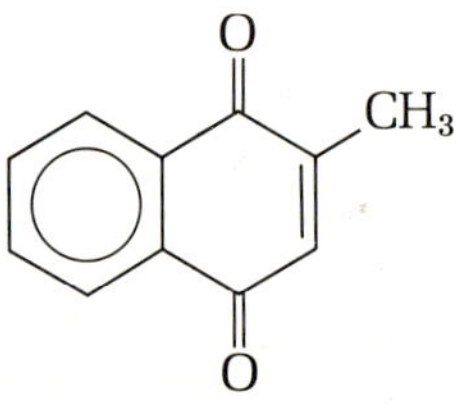

MENADIONE
(PROMOTES BLOOD CLOTTING)

Fused rings

Fused rings are rings that share a pair of atoms as in cholesterol, the sex hormones, and oral contraceptives which are known collectively as *steroids*. Their molecules all contain the following "carbon-skeleton," to which various groups of atoms are attached:

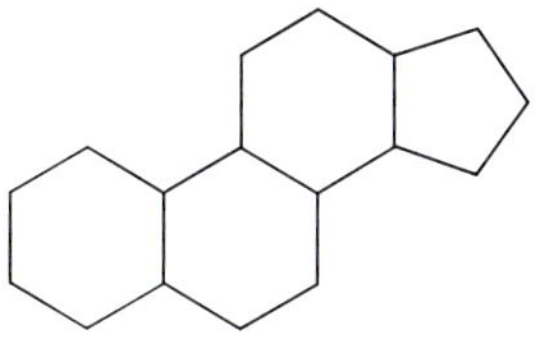

Remember that carbon atoms are at the corners of the rings.

Rings with carbon and other atoms

Rings that contain other atoms as well as carbon are of considerable biological importance. Examples are glucose and penicillin with the following molecular skeletons:

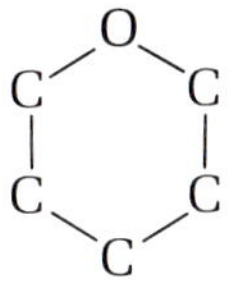

GLUCOSE
SKELETON

which may be abbreviated to

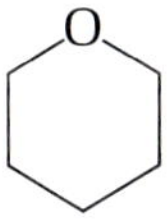

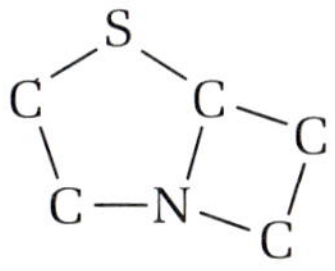

PENICILLIN
SKELETON

which may be abbreviated to

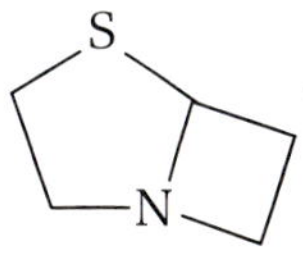

Compounds whose molecules have rings containing other atoms as well as carbon are called *heterocyclic compounds*. If the rings contain carbon atoms only, the compounds are said to be *homocyclic*.

21

Shapes of Organic Molecules

In Chapter 20, we considered some of the many ways in which carbon atoms can join together; let us now turn our attention to the question of molecular shape, for this plays an important part in determining the properties of a substance. The shapes of many molecules may be established by sophisticated experimental techniques such as X-ray diffraction and electron diffraction. A description of these techniques is beyond the scope of this book. However, we can predict the shapes of some of the smaller molecules by simple reasoning based upon the way the atoms join together. The predicted shapes of the following molecules have all been confirmed by physical measurement using the above techniques.

MOLECULAR SHAPES OF SPECIFIC COMPOUNDS

Methane, CH_4

The carbon atom is situated at the center of this molecule, surrounded by the four hydrogens. This may be represented as

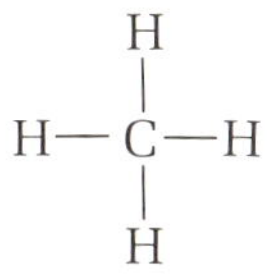

This formula implies that the bond angles are each 90°, but if we try to construct a physical picture we will see that this is not actually the case. We realize that the electron-pair bonds clearly repel each other because the electrons have the same charge. The bonds will therefore want to be as far apart as is physically possible. Thus, if the molecules were flat, or two-dimensional in shape, we would conclude that the maximum bond angle would be 90° as we first thought by just looking at the formula. However, the molecule is a three-dimensional structure having maximum bond angles of 109.5°. This structure is arrived at by the hydrogen atoms arranging themselves as though they were at the corners of a regular tetrahedron (or a triangular-based pyramid) with the carbon atom situated at the center, as shown in Fig. 21.1.

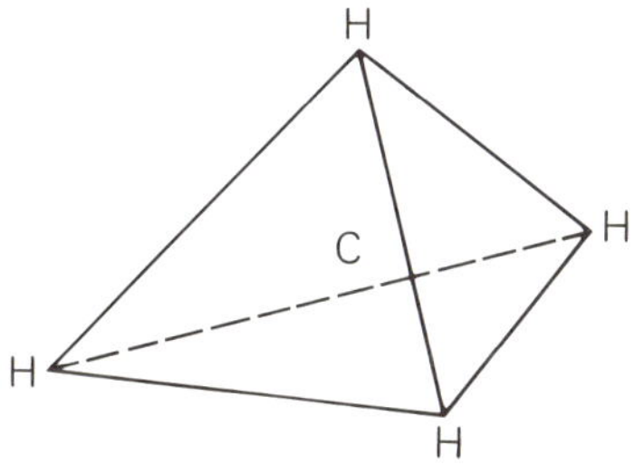

Figure 21.1
Methane showing hydrogen atoms forming a tetrahedron.

The bonds existing between the carbon atom and the hydrogen atoms have been omitted for clarity in Fig. 21.1. However, the arrangement of these bonds is illustrated in the ball-and-spoke model of methane (Fig. 21.2(a)). It is not a scale model; its main function is to represent the existence of the bond angles and the relative positions of the atoms in space. In an actual molecule the atoms are much closer together. In fact, in order to share electrons to form covalent bonds, the outer shells must overlap. Methane is therefore more realistically represented by what is called a "space-filling"model, Fig. 21.2(b). This type of model, of course, does not show the bond angles as clearly as a ball-and-spoke model does.

We can indicate the tetrahedral disposition of the bonds by writing the formula of methane like this

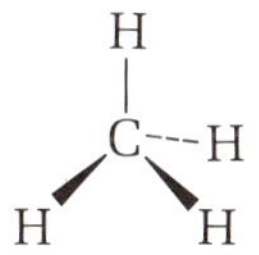

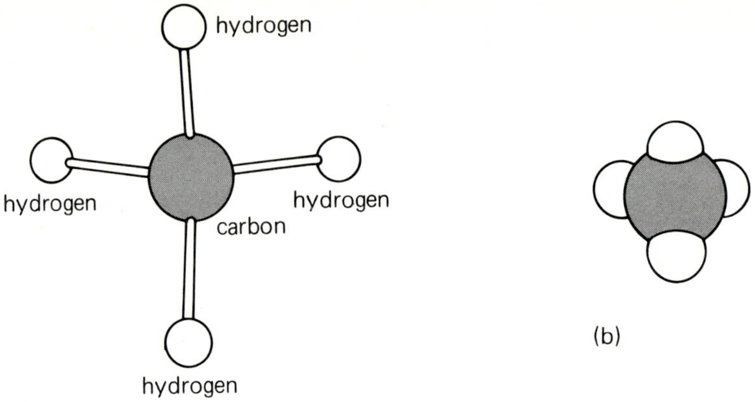

(a)

Figure 21.2
Methane: three dimensional shape represented by (a) ball and spoke and by (b) space-filling model.

By convention, the wedge-shaped bonds are understood to come out of the paper and the dotted bond to go behind, or into, the paper. The fourth bond, written normally, lies in the plane of the paper. With practice at looking at this type or formula we can visualize the three-dimensional shape of the molecule quite readily.

Chloroform or trichloromethane, $CHCl_3$

This molecule, like that of methane, has a central carbon atom with four bonds attached. It, too, will therefore have a tetrahedral or pyramidal shape.

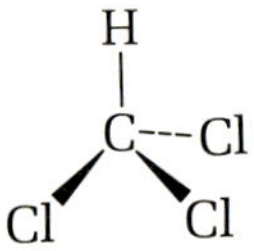

However, in this case not all of the surrounding atoms are the same size. A hydrogen atom is much smaller than a chlorine atom. The bond angles will therefore differ slightly from the 109.5° bond angles of the methane structure.

Carbon tetrachloride or tetrachloromethane, CCl_4

This molecule will have a regular tetrahedral shape because all the atoms surrounding the carbon atom are the same size.

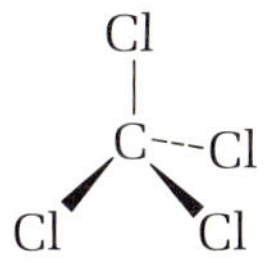

Cyclohexane, C_6H_{12}

As its name suggests, and as we have already seen in Chapter 20, cyclohexane has a cyclic or ring structure composed of six carbon atoms (Gr. *hex*, six). Since carbon's four single bonds are tetrahedrally disposed, we know that the cyclohexane molecule is not flat. The two most stable shapes or conformations that are known to exist are

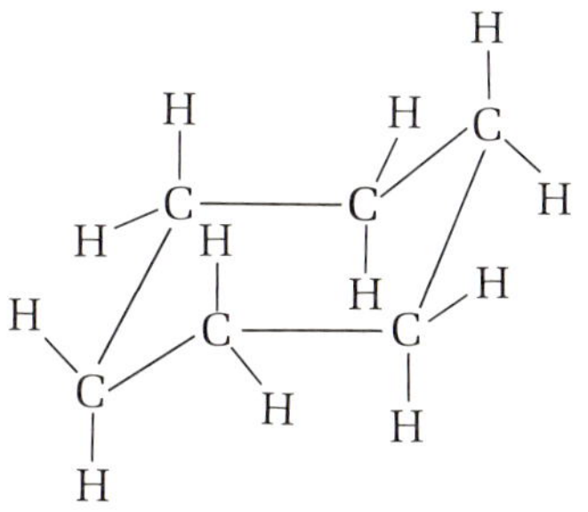

abbreviated to

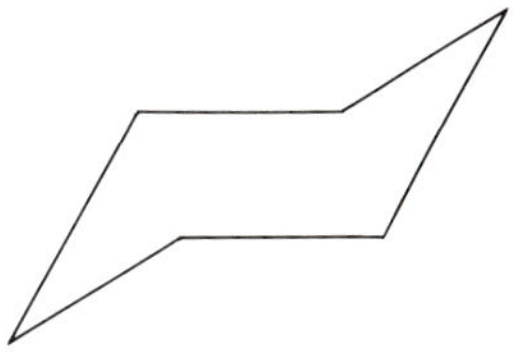

CHAIR CONFORMATION

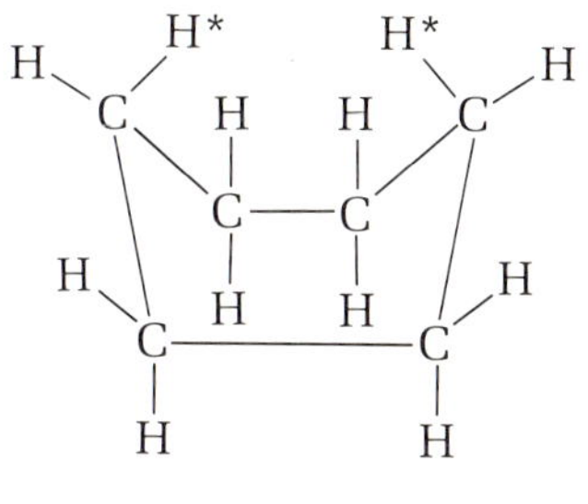

abbreviated to

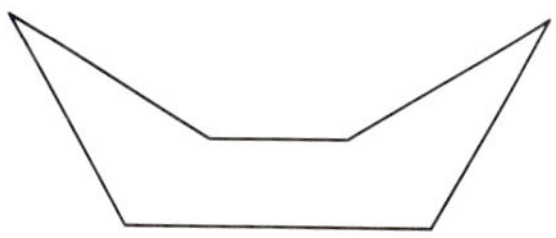

BOAT CONFORMATION

Of these, the chair conformation is the more stable structure because the asterisked hydrogen atoms in the boat form get in each other's way, this is known as *Steric hindrance*. Consequently, most of the molecules of cyclohexane are in the chair form.

The glucose "skeleton" also has this shape:

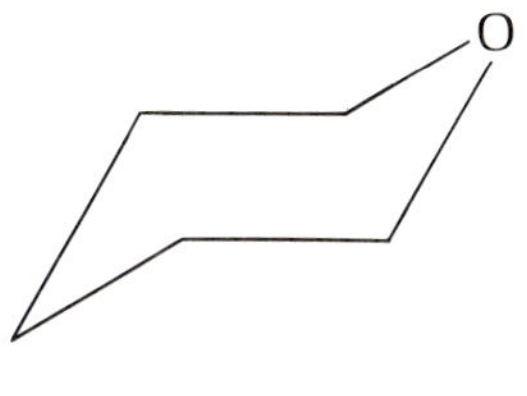

The steroid skeleton which may be represented two-dimensionally as

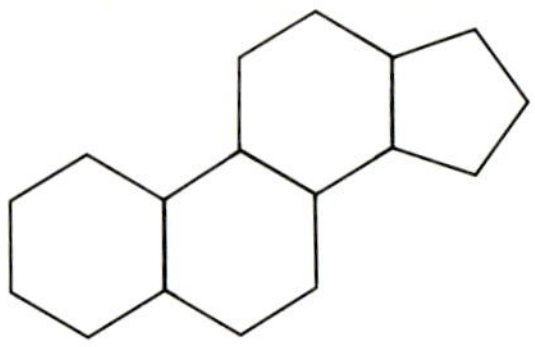

has the following three dimensional shape:

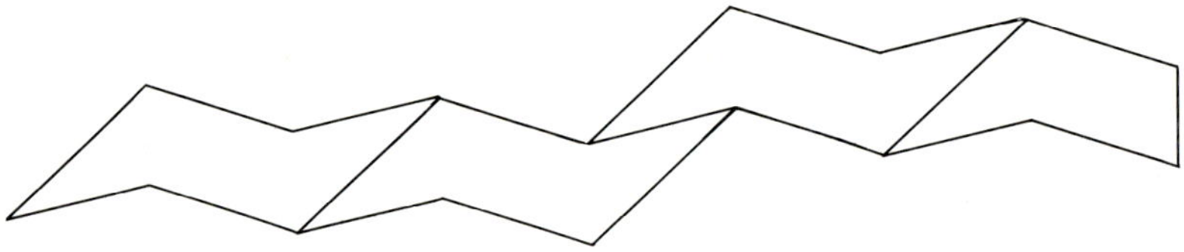

Formaldehyde or methanal, CH_2O

In this molecule, carbon and oxygen atoms are united by a double bond (a group of four electrons).

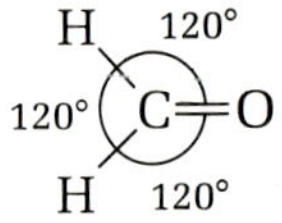

Inspection of the above formula shows that around the carbon atom there are three groups of electrons: two groups of two electrons (the two single bonds) and one group of four electrons (the double bond). These groups repel each other and therefore attempt to be as far apart as possible. They can achieve maximum separation if the molecule is flat and the bond angles are 120°.

By the same reasoning,

whenever two single and one double bond are the only bonds attached to an atom they will be coplanar (in the same plane) and the bond angles will be about 120°.

The precise angles will depend on the sizes of the atoms or groups of atoms which are attached to the central atom.

Acetylene or ethyne, C_2H_2

The carbon atoms in this molecule are united by a triple bond (a group of six electrons).

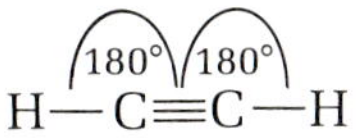

Each carbon atom therefore has two groups of electrons around it; one group of two electrons (the single bond) and one group of six electrons (the triple bond). The maximum separation of these bonds is achieved when the bond angle is 180°. The molecule of acetylene is therefore linear (or straight).

A single and a triple bond on the same atom (if they are the only bonds) are always 180° apart.

Benzene, C_6H_6

The shape of the benzene molecule cannot be predicted by the simple reasoning employed in the other examples. X-Ray diffraction methods tell us that the molecule is a flat, regular hexagon with the carbon atoms at the corners, each attached to a hydrogen atom. All bond angles are 120°.

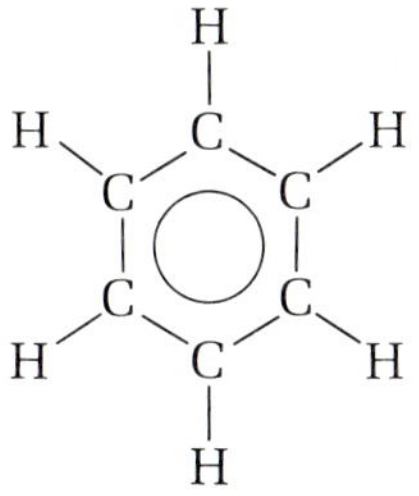

The significance of the circle has already been explained in Chapter 20. Since we know this to be the structure, we realize that the three electron-pairs around each carbon atom not only determine the bond angles but also the coplanar nature of these bonds. The molecule as a whole is planar (flat) so that the six delocalized electrons can circulate around the ring most easily.

MOLECULAR SHAPES OF ORGANIC COMPOUNDS

Before we leave the general subject of molecular shape, there is an important consequence that follows from the tetrahedral arrangement of four bonds about a carbon atom. If the four atoms (or groups of atoms) attached to a particular carbon atom are all different, then these atoms can be arranged around a given carbon atom in two different ways, and one structure will be the mirror-image of the other structure. For example, let us consider the two possible structures (or isomers) of an α-amino acid. These acids have the general formula

where R is a group of atoms, for example, an alkyl group.

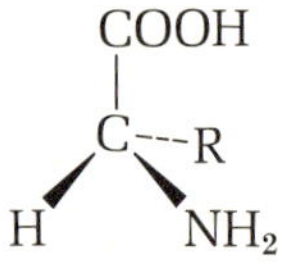

I

D-AMINO ACID

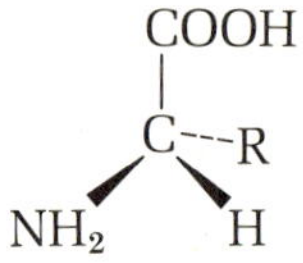

II

L-AMINO ACID

These two structures are clearly not identical because they cannot be superimposed on one another. If the carbon atom, the COOH group and the R group of structure I are placed on their counterparts in Structure II, the hydrogen atom of structure I lies on the NH_2 group of structure II, and vice versa.

These two isomers are known as *optical isomers* because they affect polarized light differently. This is, in fact, the only demonstrable difference in their properties. Exactly how and why they affect polarized light differently need not concern us here.

The arrangement of atoms and groups in isomer I is called the D-configuration and that in isomer II the L-configuration. Nature often shows a preference for one optical isomer. Thus proteins are built up exclusively from L-amino acids; none of the amino acids which nature uses in protein synthesis have the D-configuration.

FUNCTIONAL GROUPS OF ORGANIC COMPOUNDS

Now that we have some knowledge of the molecular structure of organic compounds, it is appropriate for us to consider the various atomic groups that can occur in organic molecules. These groups, by the nature of their own structure, influence the way in which the organic molecules behave or function. These groups are called *functional groups*.

Organic compounds are classified on the basis of the functional groups that they contain. Thus, compounds which contain the CHO group are called *aldehydes*. Examples of functional groups and the compounds containing them are given in Table V.

Since a functional group largely determines the behavior of a compound, it follows that compounds with the same functional group will have very similar chemical properties. A polyfunctional compound is a compound that contains more than one functional group. The properties of one functional group are often not much affected by the presence of other functional groups within the molecule. Thus, the properties of a polyfunctional compound are the sum of the properties of its functional groups.

Some important properties of certain functional groups are summarized in Table VI.

Table V. FUNCTIONAL GROUPS

FUNCTIONAL GROUP	CLASS OF COMPOUND	TYPICAL EXAMPLES FORMULA	TYPICAL EXAMPLES NAME
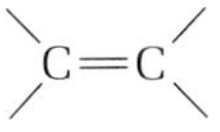	alkene	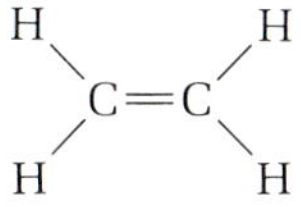	ethylene or ethene
$—C{\equiv}C—$	alkyne	$CH_3—C{\equiv}C—H$	propyne
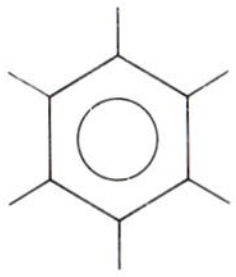	aromatic	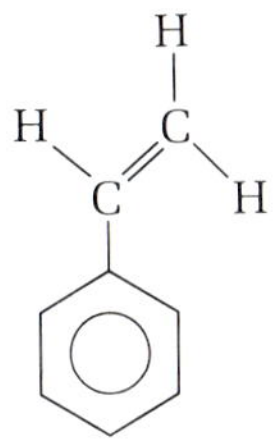	styrene
$—NH_2$	primary amine	$CH_3—NH_2$	methylamine
	secondary amine	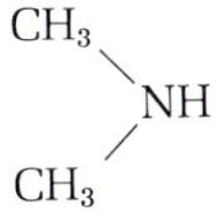	dimethylamine
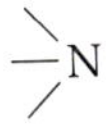	tertiary amine	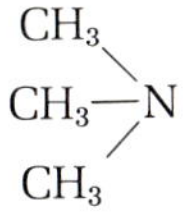	trimethylamine
—OH	alcohol	$C_2H_5—OH$	ethyl alcohol or ethanol
—OH attached to benzene ring	phenol	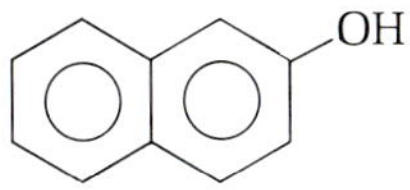	β-naphthol
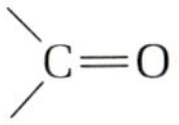	ketone	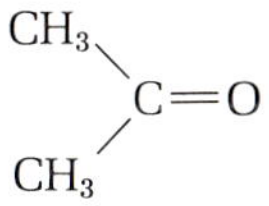	acetone

Table V. *(continued)*

FUNCTIONAL GROUP	CLASS OF COMPOUND	TYPICAL EXAMPLES FORMULA	NAME
$-C(=O)-H$	aldehyde	$CH_3-C(=O)-H$	acetaldehyde
$-C(=O)-OH$	carboxylic acid	$CH_3-C(=O)-OH$	acetic acid
$-C(=O)-NH_2$	amide	$CH_3-C(=O)-NH_2$	acetamide
$-C(=O)-O-$	ester	$CH_3-C(=O)-O-C_2H_5$	ethyl acetate
$-O-$	ether	$C_2H_5-O-C_2H_5$	diethyl ether
$-C\equiv N$	cyanide or nitrile	$CH_3-C\equiv N$	methyl cyanide or acetonitrile
$-N(=O)\rightarrow O$	nitro compound	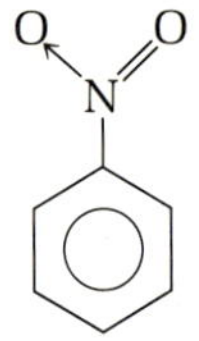	nitrobenzene
$-F$, $-Cl$, $-Br$ or $-I$	halide	C_2H_5-Cl	ethyl chloride or chloroethane
$-S(\rightarrow O)_2-NH_2$	sulfonamide	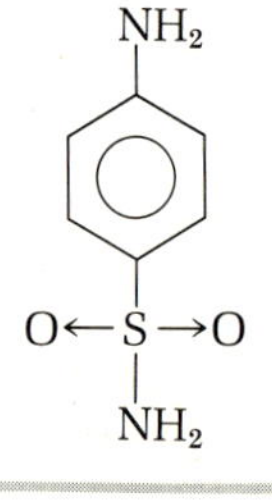	sulfanilamide

Table VI. PROPERTIES OF FUNCTIONAL GROUPS

FUNCTIONAL GROUP	IMPORTANT PROPERTIES
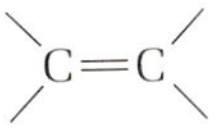	Adds bromine across double bond to give 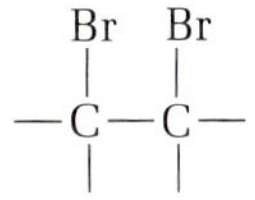
—C≡C—	Adds bromine across triple bond to give first 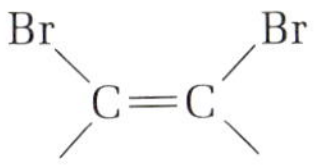 and then 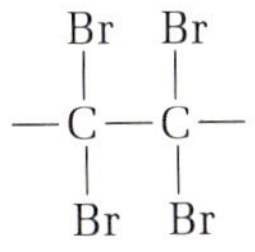
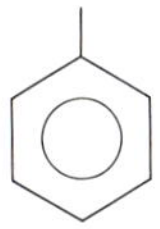 (phenyl)	Hydrogen atoms can be replaced by the groups $—NO_2$, $—SO_3H$, halogen, and alkyl. This is a characteristic of *aromatic* compounds.
$—NH_2$ (amino)	A basic group: combines with acids to form salts; for example, $—NH_3^{\oplus}Cl^{\ominus}$
—OH (phenolic hydroxyl)	An acidic group: reacts with alkalis to form salts; for example, $—O^{\ominus}Na^{\oplus}$
—OH (alcoholic hydroxyl)	A neutral group: reacts with acids to form esters; for example, 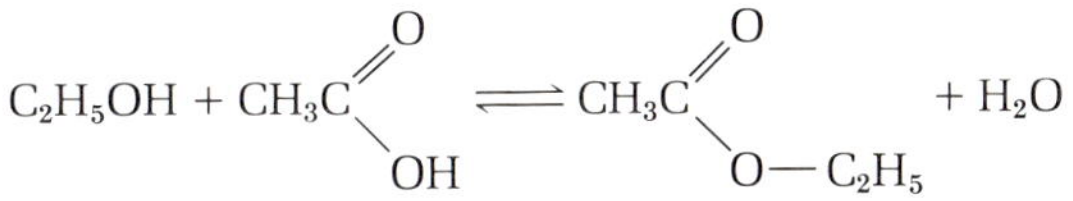 ethanol, acetic acid, ethyl acetate (ester)
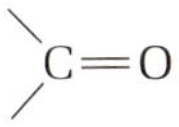 (carbonyl)	(i) Adds HCN to give 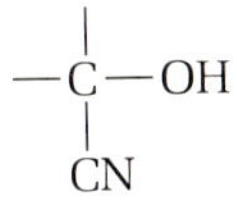 (ii) Reduces to a secondary-alcohol group 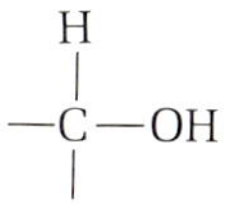

Table VI. *(continued)*

FUNCTIONAL GROUP	IMPORTANT PROPERTIES
$-C(=O)H$ (aldehyde)	(i) Adds HCN to give $-C(H)(CN)-OH$ (ii) Reduces to a primary-alcohol group $-C(H)(H)-OH$ (iii) Reduces Fehling's solution to a red precipitate of copper(I) oxide Cu_2O
$-C(=O)OH$ (carboxyl)	(i) An acidic group: reacts with alkalis to form salts; for example, $-C(=O)O^{\ominus}K^{\oplus}$ (ii) Reacts with alcohols to form esters.
$-C(=O)NH_2$ (amide)	(i) Reacts with hot alkali solution to give off ammonia. (ii) Reacts with alkali and bromine to remove $>C=O$ group leaving the $-NH_2$ group; forms a primary amine.
halogen	(i) Reacts with hot aqueous alkali to form an alcohol; for example, $C_2H_5Br + OH^{\ominus} \longrightarrow C_2H_5OH + Br^{\ominus}$ (ii) Reacts with hot alcoholic alkali to form an alkene; for example, $C_2H_5Br + OH^{\ominus} \longrightarrow H_2C{=}CH_2 + H_2O + Br^{\ominus}$ (iii) Reacts with hot potassium cyanide KCN to form a cyanide; for example, $CH_3I + KCN \longrightarrow CH_3CN + KI$
$-C{\equiv}N$ (cyanide)	(i) Reduces to $-CH_2-NH_2$; a primary amine (ii) Reacts with acidified water to give a carboxylic acid $-C(=O)OH$ For example, $CH_3-C{\equiv}N + 2H_2O + H^{\oplus} \longrightarrow CH_3-C(=O)OH + NH_4^{\oplus}$ methyl cyanide → acetic acid

22

Sugars and Starches

Sugars and starches belong to the class of compounds called the carbohydrates. They are composed of carbon, hydrogen and oxygen; the last two elements always being in the same proportions as they are in water.

The carbohydrates provide an important source of energy for humans and animals. Some are very specialized for certain jobs, for example, lactose in milk for the nourishment of newborn babies. Or they may form a vital part of very complex molecules, for example, ribose in RNA, ribonucleic acid which is a key substance for transmitting the information that controls cell function.

Carbohydrates can be divided into four groups:

1. *Monosaccharides* – simple sugars.
2. *Disaccharides* – two simple sugars joined together (with the loss of a water molecule).
3. *Oligosaccharides* – two to ten simple sugars joined together.
4. *Polysaccharides* – more than ten simple sugars joined together.

MONOSACCHARIDES (SIMPLE SUGARS)

The monosaccharides are the simplest units from which the more complex sugars are built. Probably the best known simple sugar is glucose, the sugar referred to in blood sugar. Since it has six carbon atoms, it is called a hexose, and its formula is $C_6H_{12}O_6$.

Glucose is a sugar that can immediately be used by the tissue cells as a source of energy. When taken orally, it is very rapidly absorbed, ideal for someone who is "faint for lack of nourishment." It forms the basis of several medical pep drinks. Sterile solutions of glucose can be injected intravenously for the treatment of coma due to low blood sugar (hypoglycemia) from excess insulin. The normal fasting blood glucose is 3.6 to 5.6 mmol/1 (65 to 100 mg/100 cm^3). It is raised in diabetes mellitus.

The smallest monosaccharide molecules are the trioses with three carbon atoms. The tetroses have four carbon atoms; the pentoses have five; the hexoses six and the heptoses seven. All the simple sugars have a similar pattern to their formulas.

SUGAR	FORMULA	EXAMPLE
trioses	$C_3H_6O_3$	glycerose
tetroses	$C_4H_8O_4$	erythrose
pentoses	$C_5H_{10}O_5$	ribose
hexoses	$C_6H_{12}O_6$	glucose and fructose
heptoses	$C_7H_{14}O_7$	sedoheptulose

From this table it can be seen that the general formula for the monosaccharides is $C_nH_{2n}O_n$.

All the monosaccharides and most of the disaccharides, with the notable exception of sucrose, are reducing sugars, that is, they reduce Fehling's (and Benedict's) solution to give a red precipitate of copper(I) oxide, Cu_2O. These reactions provide a means of detecting and measuring the sugar in the urine of a diabetic.

Glucose is by far the most important monosaccharide. It is one of Nature's favorite building bricks. For instance, the gigantic molecules of cellulose, starch and glycogen are all built-up from thousands of glucose molecules. We note that compounds, composed of such enormous molecules as these, are called *polymers* (Gr. *poly*, many + Gr. *meros*, part). The term *polysaccharide* is used for carbohydrates whose molecules consist of very many simple sugar molecules. The glucose molecule contains the following 'puckered' ring of atoms called the pyranose ring, which is usually represented as

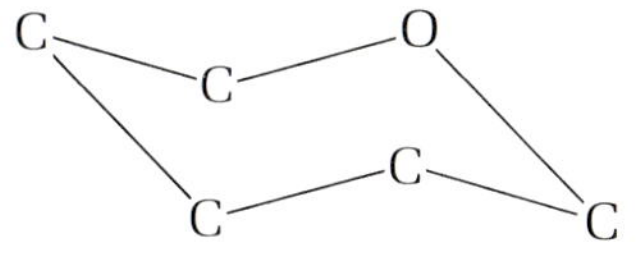

PYRANOSE RING

Alternative ways of writing this skeletal structure are

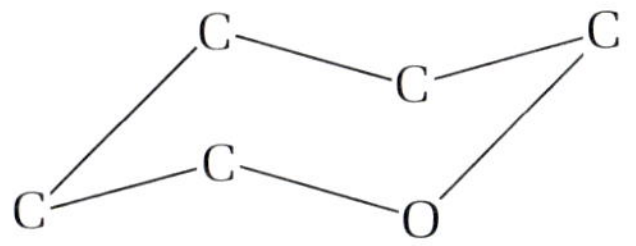

and

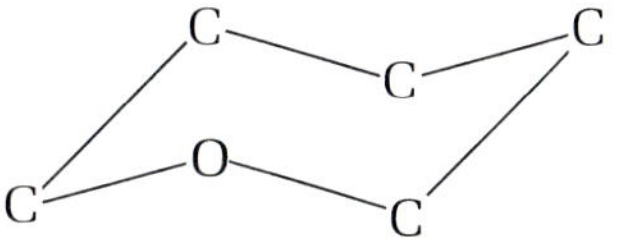

If we add the other atomic groupings and leave out the ring carbons for clarity, we have

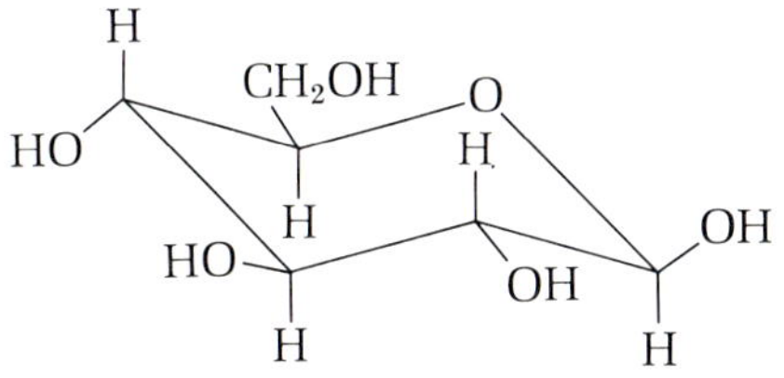

GLUCOSE

There are many other sugars which possess this pyranose ring and which also have the same molecular formula, $C_6H_{12}O_6$, as glucose. They differ from each other only in the way that the H and OH are attached to the ring. Galactose, which is combined with glucose in lactose (see below), has the following structure:

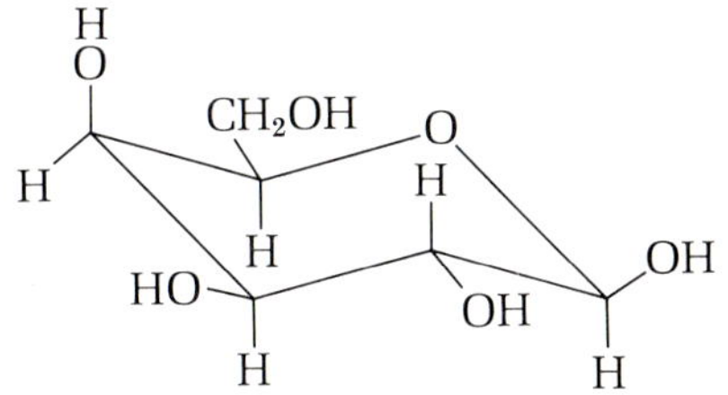

GALACTOSE

This may seem exactly the same to you as the glucose structure. However, if you look carefully, you will observe that the OH group on the left of the molecule is pointing upwards in galactose and downwards in glucose. A subtle difference surely, but Nature has no difficulty in spotting it, even if we do.

DISACCHARIDES

Disaccharides are sugars constructed from two monosaccaride units. The best known example is sucrose, the sugar which we usually have on our breakfast table. The formula for sucrose is $C_{12}H_{22}O_{11}$. Another important disaccharide with the same molecular formula is lactose, the sugar present in the milk of humans, cows and other mammals. When two molecules of a monosaccharide are joined together to form a disaccharide, one molecule of water is eliminated

$$2C_6H_{12}O_6 = C_{12}H_{22}O_{11} + H_2O$$

So the general formula for a disaccharide is $C_n(H_2O)_{n-1}$.

Lactose

The lactose molecule is derived from a glucose and a galactose molecule, joined together by an oxygen-bridge. Its skeletal structure may be represented like this

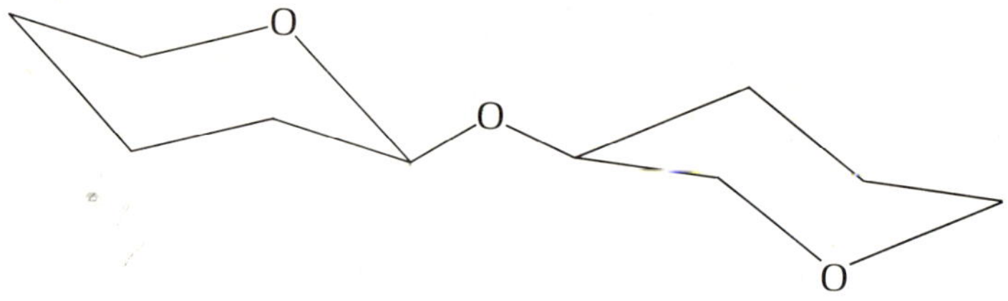

GALACTOSE UNIT GLUCOSE UNIT

LACTOSE

Sucrose

Sucrose differs from lactose in having a fructose instead of a galactose unit. Fructose, when forming part of the sucrose molecule, has a furanose ring "skeleton."

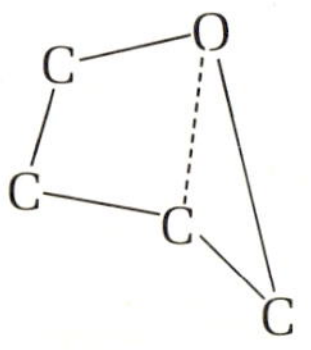

FURANOSE RING

The dotted lines are merely to indicate the shape of the molecule, namely a quadrilateral with a triangular flap turned down.

A fuller structure is

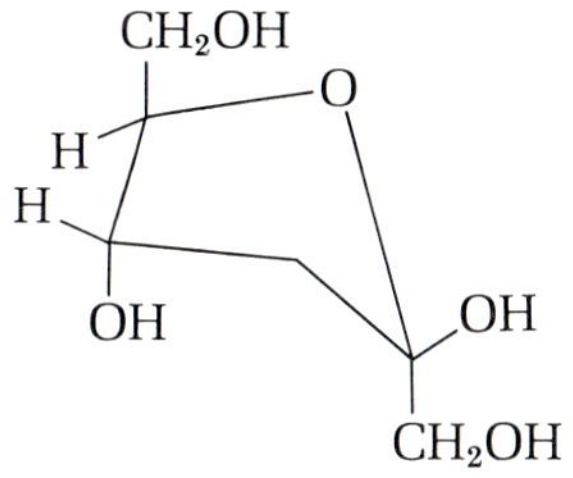

FRUCTOSE

In the sucrose molecule, the glucose and fructose units are joined together in the manner illustrated by the following skeletal structure:

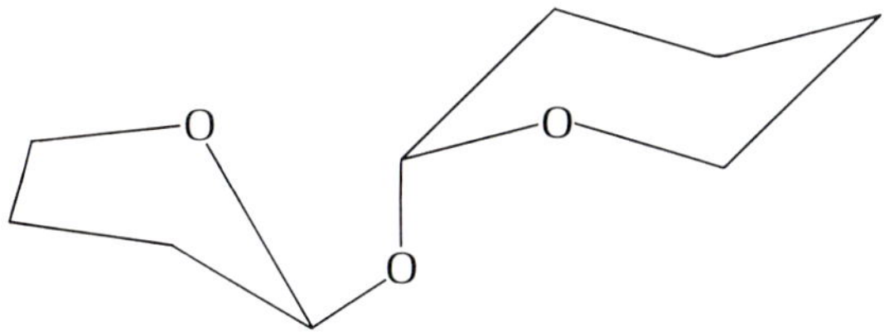

FRUCTOSE UNIT GLUCOSE UNIT

SUCROSE

The commercial production of sucrose probably exceeds that of any other pure organic chemical.

Human consumption of lactose and sucrose

When lactose and sucrose are taken orally they are digested by enzymes in the alimentary tract which add a molecule of water and at the same time split them into their component simple sugars namely,

$$H_2O + C_{12}H_{22}O_{11} = 2C_6H_{12}O_6$$

This process of adding water and splitting is called *hydrolysis* (Gr. *hydro*, water + Gr. *lysis*, loosing). Once the disaccharides have been hydrolyzed, the simple sugars can be readily absorbed.

POLYSACCHARIDES

The two most important, and best known, polysaccharides are cellulose and starch. Both are built-up from glucose units and have the general formula $(C_6H_{10}O_5)_n$. They differ only in the way the glucose units are joined together. Cellulose is the structural material of plants, giving them their shape and rigidity. It is probably the most widespread of all the organic compounds. Starches provide plants with a valuable means of storing energy; they make up their reserve food supply and are stored mainly in the seeds. Man relies heavily on both these polysaccharides. Cellulose is used as timber for constructional purposes, as cotton and rayon for clothing, as paper for communication and packaging and as absorbent material for surgical dressings. Starch in the form of potatoes, rice, wheat, corn, and so forth supplies a large part of our food and energy requirements. It is hydrolyzed by enzymes, such as amylase (ptyalin) in the saliva into carbohydrates with smaller molecules, called dextrins, which have a sweetish taste. You may have noticed this if you have chewed a piece of bread very slowly. The dextrins are further hydrolyzed to glucose that is carried in the bloodstream to the tissues. Its oxidation, ultimately to carbon dioxide and water, supplies the tissues with the energy they require. Some of the glucose is taken to the liver where it is polymerized to glycogen (animal starch) which serves as an energy-store for us. It is stored in the liver as liver glycogen or in the muscles as glycogen. When required, the muscle glycogen is broken down into glucose again.

Most mammals, including man, lack the enzymes needed to hydrolyze cellulose so they cannot use it as a food. However, ruminants, or cud-chewing animals, such as the cow, have bacteria in their digestive tract and these bacteria obligingly break down cellulose into glucose for the animal. The glucose units in cellulose are joined together in this manner

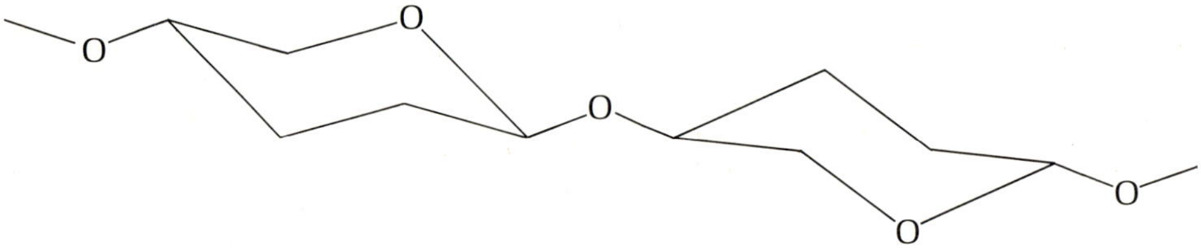

CELLULOSE

There is considerable hydrogen bonding between the chains and this gives cellulose its great mechanical strength. It also explains its insolubility, for the chains are unable to separate and mix thoroughly with water molecules.

In starch the glucose molecules are joined together in a different way. There are two components to starch: amylose, which is soluble, and amylopectin, which is not. The skeletal structure of amylose is as follows:

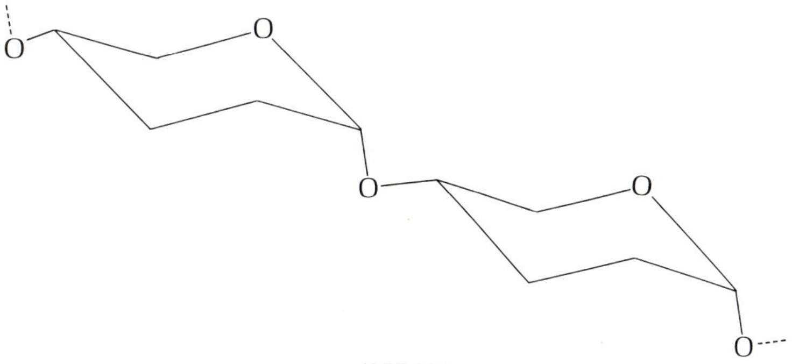

AMYLOSE

The amylose chains are believed to be coiled to form a spiral-shaped molecule.

Amylopectin differs from amylose in that some of the glucose units in one chain are attached to units in another chain giving a branched structure.

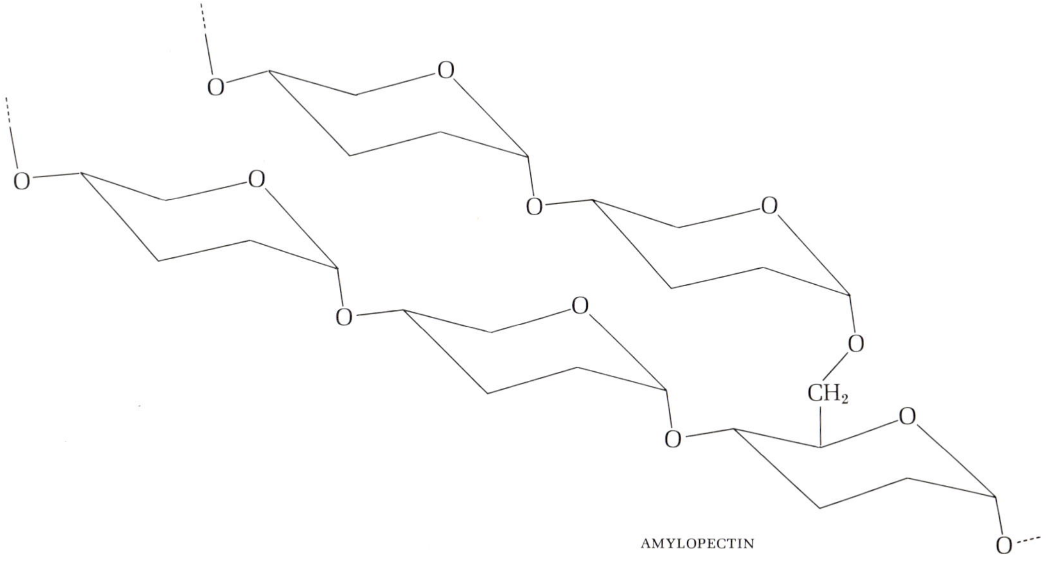

AMYLOPECTIN

The branching is more clearly shown schematically as

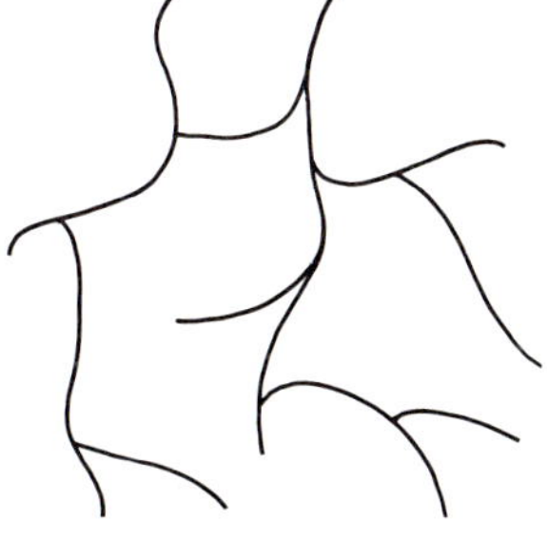

One of the effects of branching is to make the amylopectin molecules larger than those of amylose. This is why it is insoluble.

Starch contains about 20% amylose and about 80% amylopectin, the composition depending on the source of the polysaccharide. Starch is not a reducing carbohydrate and so it cannot be detected with Fehling's or Benedict's solution. It can, however, be very easily identified, even in trace amounts, by the deep blue color which its amylose component gives in iodine solution.

The structure of glycogen is similar to that of amylopectin except that its chains are shorter and more highly branched.

AMINO SUGARS

These can be considered as being derived from sugars by replacing an hydroxyl group, OH by an amino group, NH_2. Replacement of an OH group in glucose by NH_2 gives glucosamine. Galactose gives galactosamine. The anticoagulant, heparin, which is used in the treatment of thrombosis, is a polymer built up from glucosamine units. In the body it is produced in special cells, called mast cells, and assists in preventing blood from clotting in the blood vessels. Heart valves, cartilage, skin, adult bones, tendons, and cornea all contain the polymer chondroitin, which is composed of galactosamine units.

For some years now, polymer chemists have been imitating nature and building large molecules from small ones. In this way they make plastics, synthetic fibers, and rubbers. By using a variety of small molecules and by varying the length of the polymer chains and the degree of "crosslinking" between them, they are able to produce new materials with a wide variety of properties and uses, for example, polyethylene.

23

Fats

Fats are one of the three main classes of foods, the other two being carbohydrates (Chapter 22) and proteins (Chapter 24). They are the body's major long-term energy reserve, being called upon in times of starvation and great exhaustion. Part of the body's fat is formed directly from fat that is eaten, the remainder is produced by the body from other fats, carbohydrates and proteins in the diet. The main reservoir of fat is the adipose tissue beneath the skin. Because fat is a poor conductor of heat, the adipose tissue acts as an insulating jacket, conserving the heat of the body.

CHEMICAL STRUCTURE

Chemically, we can consider fats as being derived from the alcohol glycerol (or glycerine) and carboxylic acids. A carboxylic acid is an acid that contains a carboxyl group, COOH. The organic compound formed when an alcohol reacts with an acid is called an *ester*. For this reason fats are said to be glycerol esters, or glycerides. If we represent a carboxylic acid by the formula RCOOH, where R is a group of carbon and hydrogen atoms, then the formation of a fat can be represented by the following equation:

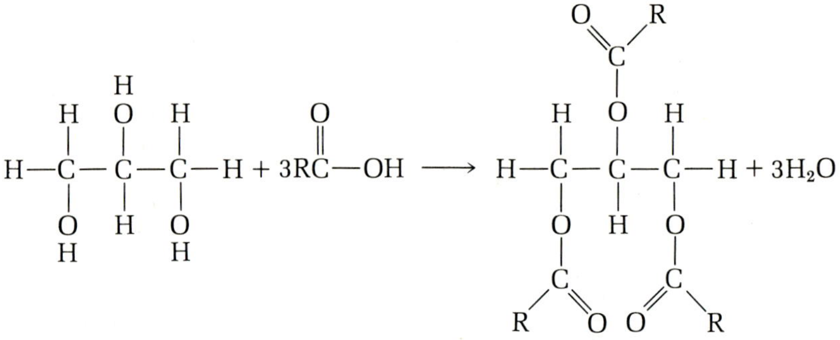

GLYCEROL | CARBOXYLIC ACID | FAT (GLYCERIDE)

The R groups commonly contain 15 or 17 carbon atoms, although in milk fats they contain considerably fewer. If there are double bonds in the R groups, the fat will normally be liquid, that, is, an oil, at room temperature. Common examples of liquid fats are olive oil, castor oil, linseed oil, and corn oil.

Vegetable oils (unsaturated fats) can be "hardened," that is, combined with hydrogen to produce solid (saturated) fats such as margarine. All that happens in this process is that hydrogen is added across the double bond.

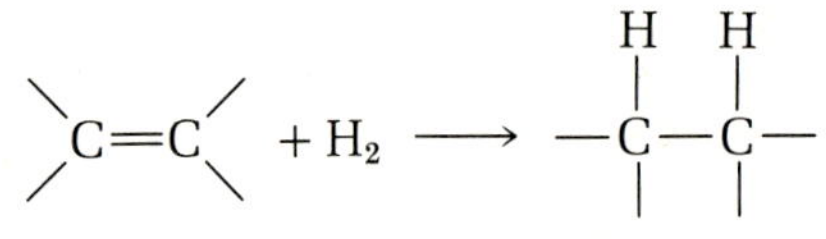

UNSATURATED | SATURATED

We see that the product of this reaction does not contain a double bond. Such fats are called saturated and are solids.

Soap and glycerol

Another important use of oils, and solid fats, is the manufacture of soap and glycerol. The fats are heated with a solution of sodium hydroxide or potassium hydroxide. The following reaction, known as saponification, takes place.

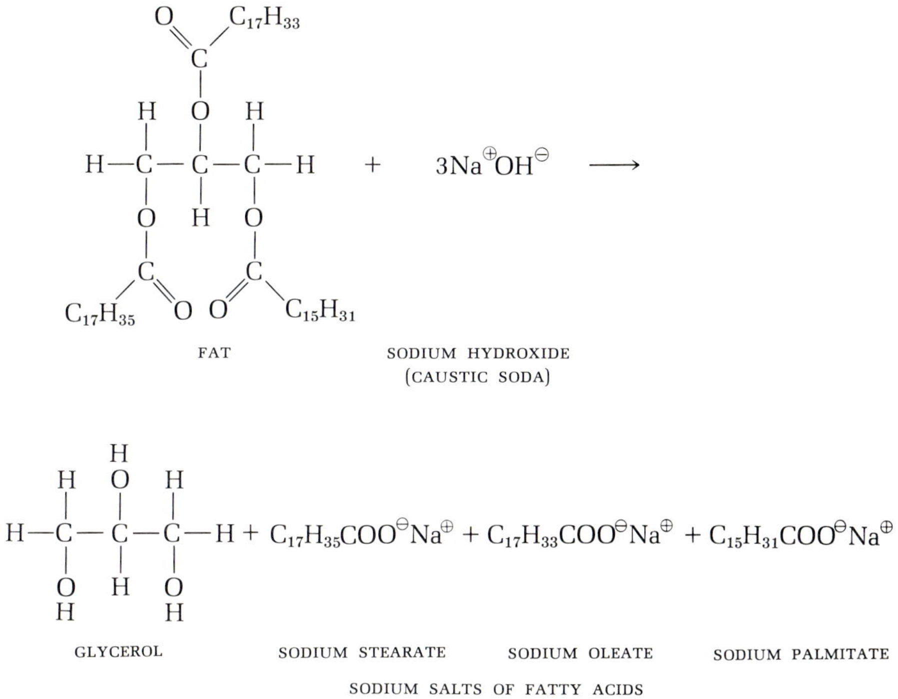

Thus, in general we have the word equation.

fat + alkali ⟶ glycerol + soap

You will notice that there are three different R groups in the fat molecule. This is usually the case. The mixture of sodium salts which is produced is called soap. It is separated from the glycerol by adding salt (sodium chloride) which causes the soap to rise to the surface where it can be skimmed off. So-called hard soaps contain the sodium salts of carboxylic acids, while soft soaps contain the potassium salts.

CLEANSING ACTION OF SOAP

How does soap clean? Our skin gets dirty because particles of dirt adhere to a thin layer of grease or oil which covers the skin. To remove the dirt, we have to remove the oil. Water alone is not effective because the oil will not dissolve. We must use something that will disperse the oil in water. Soap does this, but how? Let us look at the formula of sodium palmitate, a common constituent of soaps.

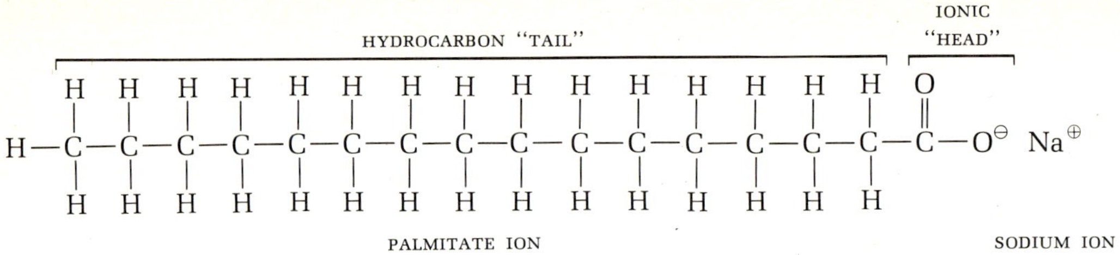

The palmitate ion in sodium palmitate has a long hydrocarbon tail, $C_{15}H_{31}$ which is soluble in oil, and a short ionic head, $COO^{\ominus}$, which is soluble in water. These so-called tails attach themselves to oil droplets which are produced by the washing action. We see in Fig. 23.1 that the droplets now have an outer coating of negative ions which dissolve in water. This negative charge on all the droplets prevents them from coming together, since like charges repel. They are therefore suspended, together with their entrapped dirt particles, in the water. A suspension of this type is called an "emulsion." The detergent or cleaning action of soap is thus due to its emulsifying ability.

Soapless detergents work on the same principle; they all consist of a long, hydrocarbon tail which is oil-soluble and an ionic head which is water soluble.

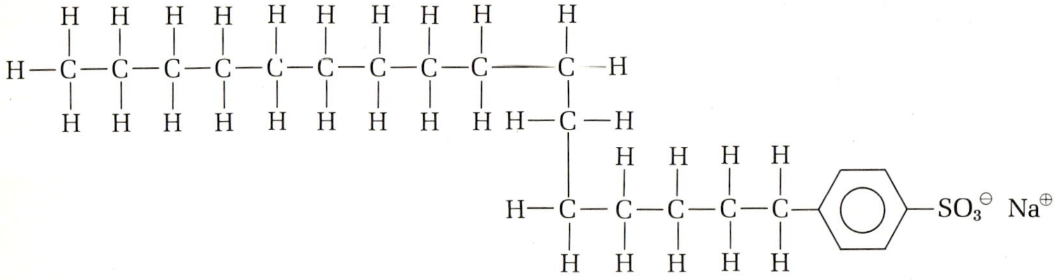

$RCOO^{\ominus}$ ions in soap combine with the ions of calcium, magnesium or iron, which may be present in water, forming a precipitate or scum (the bathtub ring.). Soapless detergents do not suffer from this disadvantage because their negative ions do not combine with the ions of these three metals.

DIGESTION OF FATS

For fats to be digested they must be emulsified in the digestive tract. The emulsifying agents or detergents are the bile salts, whose molecular structures have the same characteristic features as those of soap and soapless detergents. Once emulsified, fats are hydrolyzed to glycerol and fatty acids.

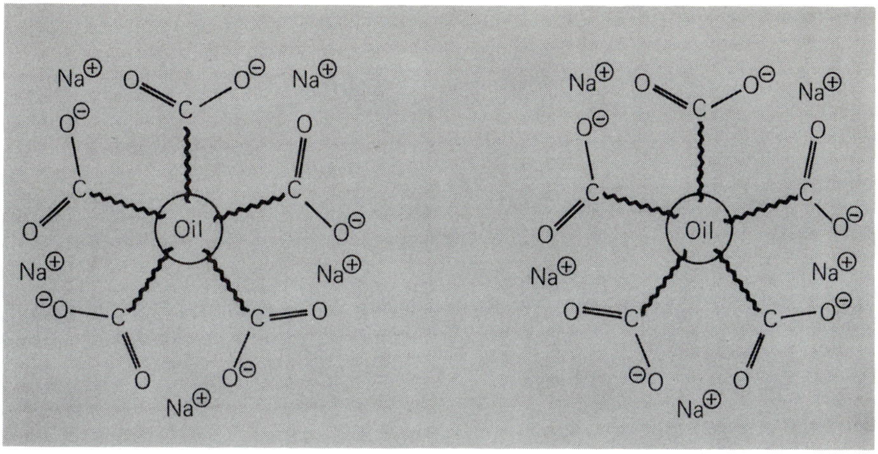

Figure 23.1
Emulsion of oil in water.

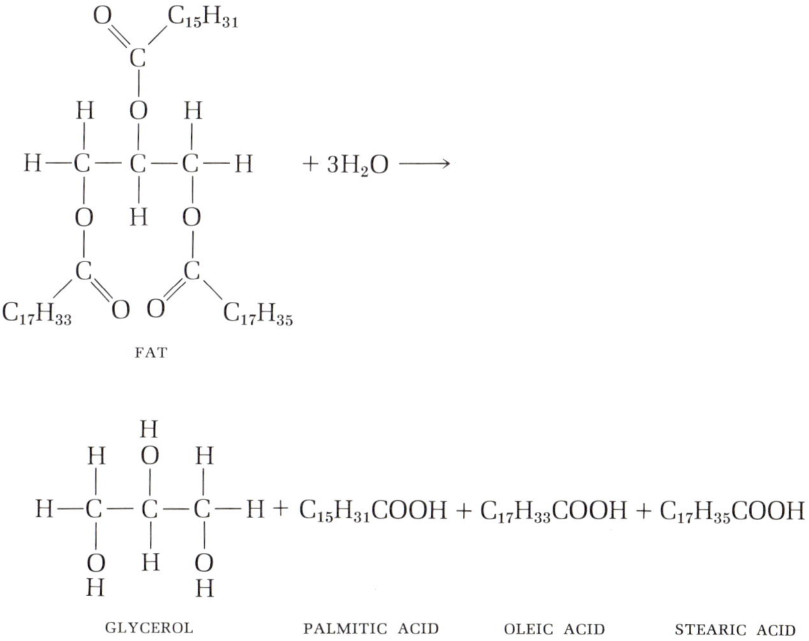

This hydrolysis is catalyzed by the pancreatic enzyme, lipase.

EXTRACTION OF FATS

Fats belong to a larger class of compounds called the lipids. These include a variety of structural types and the only reason that they are

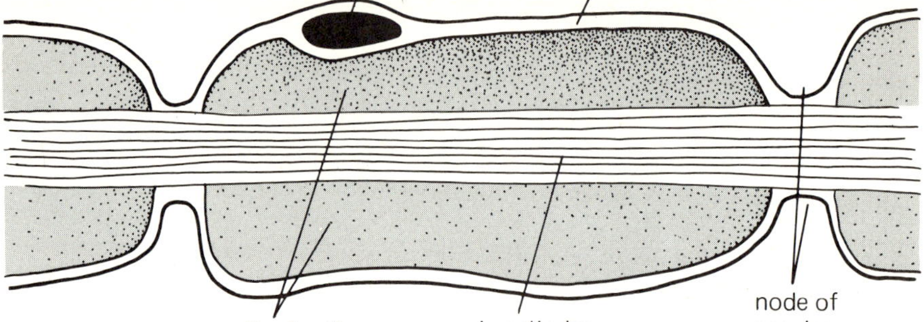

Figure 23.2
Nerve fiber: conducting tissue surrounded by insulating sheath of myelin.

grouped together is that they can all be extracted from plant and animal tissue with ether or chloroform, which dissolve them. The phospholipids are an important group of lipids and, as their name suggests, they contain phosphorus. One class of phospholipid, namely, the phosphatides, is structurally similar to the fats. Phosphatides are found in cell walls and in the membranes of cell nuclei and mitochondria. They are also believed to be involved in blood clotting.

Sphingolipids are a class of lipids occurring in the brain and nerve tissue. They are an important constituent of myelin found in the insulating sheath of nerve cells (Fig. 23.2). Myelin is extremely stable and protects the signal-bearing cores of the nerve cells. Sphingolipids contain longer fatty acid R groups than fats and these long R groups twist around one another so that the molecules are tightly bound. This contributes greatly to the strength of the nerve sheath. If, for some reason, the R groups are too short to give adequate intertwining and strengthening of the myelin sheath, the nervous system is poorly protected and conditions such as multiple sclerosis, infantile Gaucher's disease, and forms of leukodystrophy result.

24

Proteins

Proteins are very aptly named (Gr. *proteios*, first) for they are of the first importance in life processes, occurring as they do in all living cells. They are the chief constituent of muscle, tendons, skin, hair, nails, nerves, and blood. Many hormones and all the enzymes and antibodies are proteins. Like starch and cellulose, they are polymers.

STRUCTURE

The molecules of proteins are built up from small α-amino acid molecules. In the following formula R can be a hydrogen atom or one of about twenty different groups of atoms:

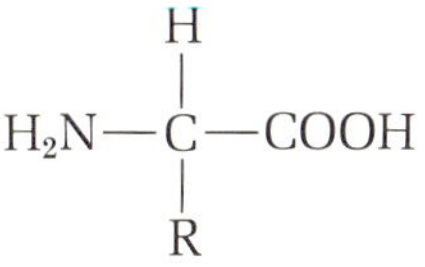

A protein is characterized by the sequence of its various amino acids and this sequence is determined by DNA and RNA molecules (Chapter 25). Each animal species has its own particular set of proteins, different ones for each part of each organ. So in the animal kingdom you can appreciate that there are millions of different proteins and yet nature uses only about twenty different amino acids to make them. This is possible because the number of different ways of arranging about twenty different units in such large molecules is virtually limitless. These amino acids and their standard abbreviations are listed in Table VII. Typically, there are seldom fewer than seventeen of them in any given protein. They link together because the amino group, NH_2 of one acid interacts with the carboxyl group, COOH of another. For example,

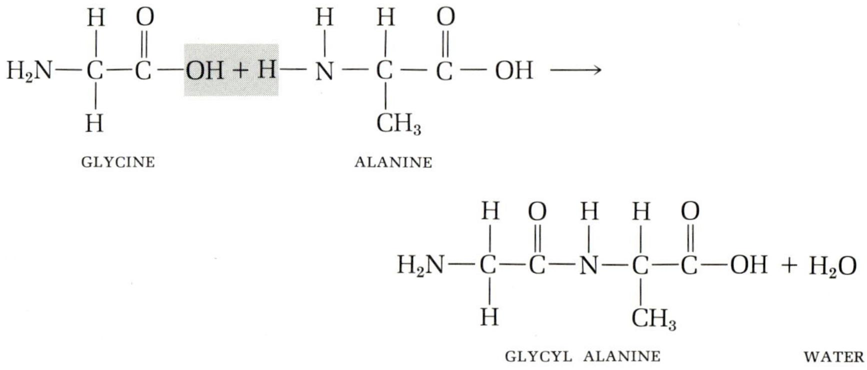

Do not be alarmed by the size of the formula of glycyl alanine because it is customarily abbreviated to Gly·Ala indicating that it contains a glycine residue and an alanine residue. Since amino acids each

Table VII. AMINO ACIDS FOUND IN PROTEINS

NAME	ABBREVIATION	NAME	ABBREVIATION
alanine	Ala	isoleucine*	Ileu
arginine	Arg	leucine*	Leu
asparagine	Asp-NH_2	lysine*	Lys
aspartic acid	Asp	methionine*	Met
cysteine	CySH	phenylalanine*	Phe
cystine	CyS-SCy	proline	Pro
glutamic acid	Glu	serine	Ser
glutamine	Glu-NH_2	threonine*	Thr
glycine	Gly	tryptophan*	Try
histidine	His	tyrosine	Tyr
hydroxylysine	Hylys	valine*	Val
hydroxyproline	Hypro		

* Essential in the diet. The body is unable to synthesize them.

contain both an amino and a carboxyl group, they can link together like beads on a string.

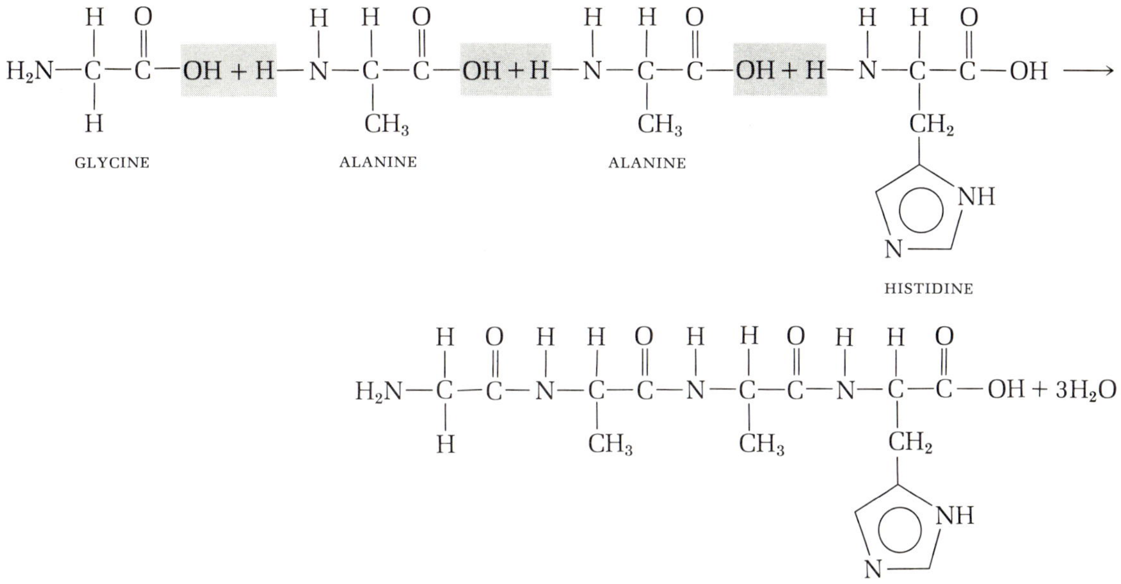

The off-putting formula of this product can be abbreviated quite simply to Gly·Ala·Ala·His. By convention, the left-hand residue contains the NH_2 group and the right-hand one the COOH group. If we look very closely at the lengthy chemical formula of Gly·Ala·Ala·His in the chemical equation, we will realize that when the amino acids are strung together they will always form the following "backbone":

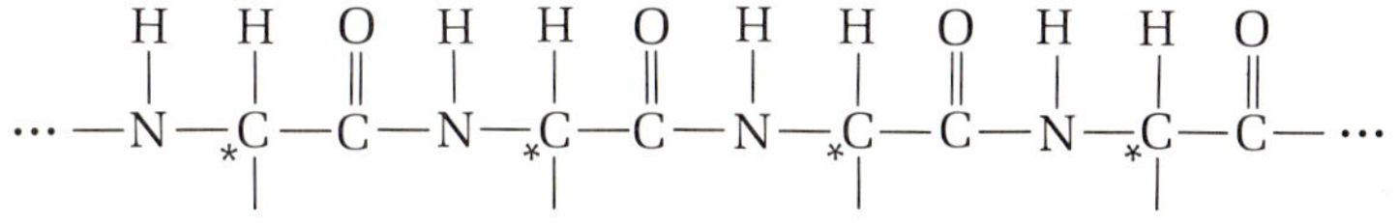

At regular intervals along this chain particular groups of atoms, the R groups of the amino acids, are attached. They are bonded to the carbon atoms marked with an asterisk. The properties of a protein depend on the nature and sequence of these R groups along the chain. Merely replacing one of them by another can produce marked and sometimes disastrous differences in behavior. The serious disease sickle-cell anemia is believed to be caused by the replacement of a glutamic acid R group by a valine R group in the hemoglobin molecule.

Proteins contain at least fifty amino acid residues. Compounds with fewer than this are called *polypeptides* or simply *peptides*. The amino acid Gly·Ala is called a dipeptide because it possesses two residues; the amino acid Gly·Ala·Ala·His possesses four residues and is called a tetrapeptide. A number of peptides play an important part in body chemistry. Thus, glutathione Glu·CySH·Gly, a tripeptide, is found in most living cells and oxytocin, a nonapeptide, with nine residues, is a posterior pituitary hormone concerned with uterine contraction.

The determination of the amino acid sequence in proteins is an exceedingly complicated business. Sanger in 1955, after ten years effort, succeeded in establishing it in insulins from various animal species, an achievement which earned him the Nobel prize for chemistry in 1958. He found that the structures of the various insulins differed only very slightly from one another. In fact, sheep insulin, whose structure is given in Fig. 24.1 can be used to treat diabetes. Some antibodies may be formed by the human recipients against it but they have little effect on the activity of the insulin. If you look at the structure of sheep insulin, you will notice that it contains the amino acid, cystine, which is, in effect, made up of two smaller amino acids.

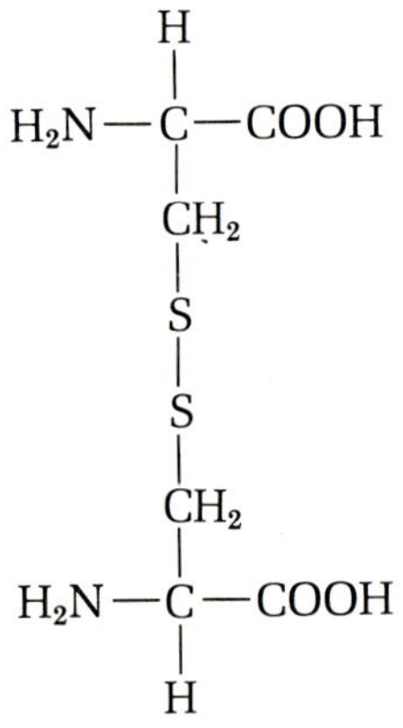

CYSTINE, CyS—SCy

This enables the two halves of it to be in two different peptide chains, the S-S group forming a bridge or crosslink between them. We notice two such crosslinks in sheep insulin. They play an important part in determining the shape of the molecule and its physiological properties.

Keratin, the protein in hair, has a large number of these crosslinks which give hair its strength. The home permanent involves breaking the S-S bonds, then setting the hair in the desired style and finally making fresh crosslinks to hold the hair in its new position.

CLASSIFICATION AND FUNCTION OF PROTEINS

There are two main classes of proteins: *fibrous proteins* and *globular proteins*. The former are insoluble in water, and the latter soluble in both water and aqueous solutions of acids, bases or salts.

Their names suggest what their molecular structures are like. Fibrous proteins consist of long, threadlike molecules. These are often held together in bundles to form a multistrand fiber by hydrogen bonds between the CO groups in one molecule and the NH groups in another.

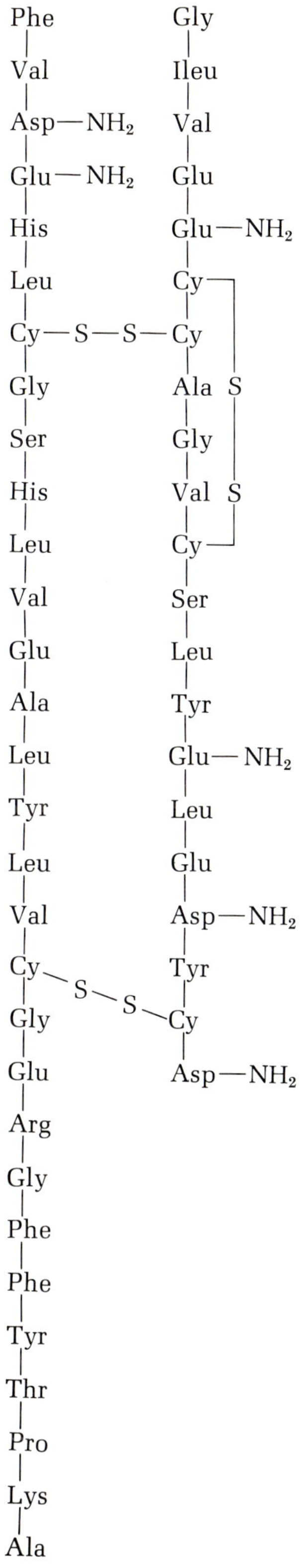

Figure 24.1
Structure of sheep insulin.

$$\overset{\delta-}{>CO}—\cdots—\overset{\delta+}{HN}<$$

They are insoluble in water because the hydrogen bonds prevent the protein molecules separating, which they must do if they are to mix with water molecules and so dissolve. The molecules of globular proteins are folded to give an approximately spherical (or globular) shape and no hydrogen bonds are formed between these molecules and therefore they are able to dissolve.

The molecular structure of fibrous proteins makes them ideal structural material for animal tissues. Myosin in muscle, collagen in tendons and keratin in skin are all fibrous proteins. Enzymes, hemoglobin and protein hormones must be soluble in order to fulfill their role in metabolism. They are, therefore, globular proteins.

CLOSER LOOK AT THE STRUCTURE OF PROTEINS

There are three main aspects to the structure of a protein. First, there is its *primary structure* which is simply the sequence of amino acids in the molecule. Secondly, there is its *secondary structure* which depends on the nature of the R groups. When these are small, parallel protein chains tend to hydrogen bond together to form pleated sheets, as in fibroin, the protein of silk. When they are large, they are best accommodated by the chain throwing itself into a spiral, known as an α-helix. Hydrogen bonding occurs between different parts of the same chain to keep it in a spiral shape. The amino acids must all have the same configuration to fit into this helix. Their configuration is such that the spiral is right-handed and not left-handed.

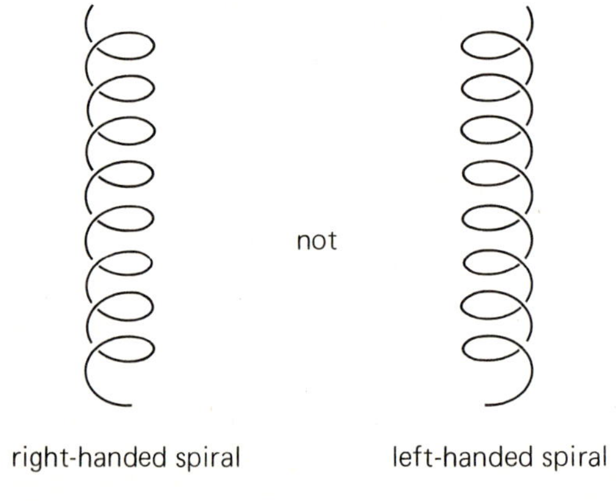

The contraction and relaxation of muscle is believed to involve the coiling and uncoiling of α-helices in myosin. Collagen, the fibrous protein of bone, tendons, skin and cornea, is not able to form an α-helix because there are insufficient hydrogen bonds available. Instead, three chains twist around each other to form a triple-stranded protein. In bone tissue, hydroxy apatite, $Ca(OH)_2 \cdot 3Ca_3(PO_4)_2$ is embedded in a matrix of collagen.

Thirdly, we note that the formation of an α-helix is not always the end of the molecular shaping process. In most fibrous proteins a *tertiary structure* is produced by α-helices twisting around each other to form "ropes." The tertiary structure of globular proteins, for example, enzymes, arises because the α-helix is folded up on itself, so that R groups which are attracted to water are on the outside of the globule and those which are not attracted are tucked away on the inside. The catalytic activity of an enzyme depends on its tertiary structure. The molecule of a reacting substance, known as the substrate, fits neatly into a cavity or cleft in the enzyme molecule (see Fig. 26.2) just as a key fits perfectly into a lock. The enzyme is then able to assist in the breakdown of the substrate. The close fit which is necessary explains the specificity of enzymes, that is, why each reaction requires a specific enzyme to catalyze it. This lock-and-key mechanism also explains the action of many drugs. A number of bacteria require, for their growth, *para*-amino benzoic acid which is present in blood and tissues. The molecule of the drug sulfanilamide is very similar in size and shape to that of para-amino benzoic acid.

SULFANILAMIDE

para-AMINO BENZOIC ACID

The drug consequently occupies cavities in the enzymes of the bacteria normally filled by *para*-amino benzoic acid. This prevents growth and multiplication of the bacteria that can then be destroyed by the body's own defense mechanisms.

The secondary and tertiary structures are very sensitive to change in the environment of the protein. They are destroyed, for example, by changes in temperature and pH, by violent shaking and by metal ions; a process known as *denaturation*. The biological action of an enzyme is therefore lost when it is denatured. Common examples of the denaturation of proteins are the clotting of egg white (albumin) when it is cooked and the formation of meringue when it is whisked. A less obvious, but no doubt equally familiar example, is the frothing of beer when it is poured. Lead and mercury, and their compounds, are toxic because the ions of these metals combine with proteins to change their structure. The denaturing ability of these ions is made use of in treating cases of lead and mercury poisoning. Egg white, a readily available protein, is administered. This combines with the metal ions, preventing them from attacking the body proteins. An emetic or gastric lavage must then be given to remove the metal-ion–albumin complex; otherwise it will be broken down again in the stomach into the free metal ion.

25

Conjugated Proteins

Some proteins form conjugated proteins by combining with nonprotein molecules called *prosthetic groups*. These groups contribute to the proteins's properties that are biologically important. For example, the prosthetic group heme is attached to the protein globin in hemoglobin. The ability of hemoglobin to combine reversibly with oxygen is dependent on the heme group with its associated iron atom. There are four heme groups in each hemoglobin molecule. Each heme group is attached to a histidine (an amino acid) residue in the globin molecule (Fig. 25.1).

The heme group consists of four pyrrole rings (Fig. 25.2) joined together as shown above to form a porphyrin. This porphyrin grouping is also found in chlorophyll and vitamin B_{12}. The enormously complicated tertiary structure of hemoglobin was determined by Perutz in 1960. He shared a Nobel prize for this work with Kendrew who had earlier elucidated the tertiary structure of the somewhat simpler conjugated protein, myoglobin which contains only one heme group in its molecule. Myoglobin stores oxygen in the muscles.

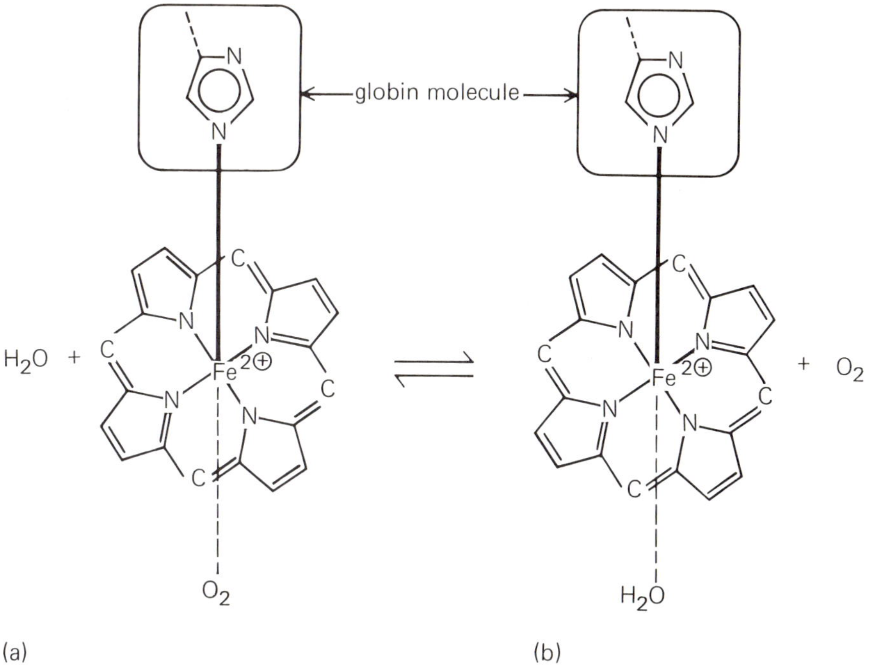

Figure 25.1
Heme configuration: (a) in oxyhemoglobin; (b) in reduced hemoglobin.

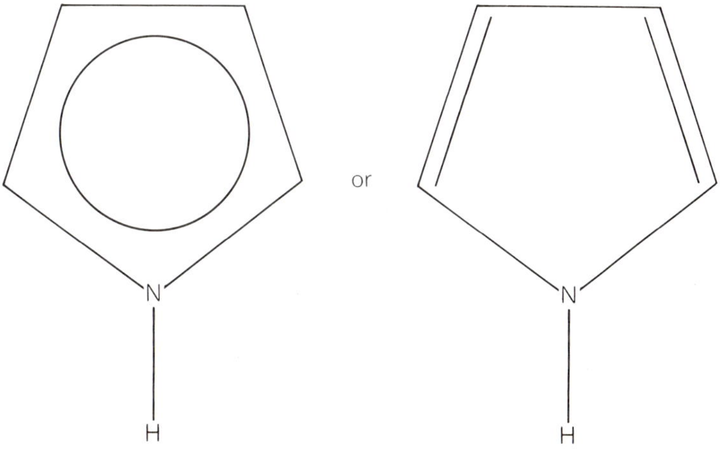

Figure 25.2
Pyrrole ring.

COENZYMES

A number of enzymes require the attachment of prosthetic groups before they can act as biological catalysts. These nonprotein groups are the chemically active sites on the molecule but they are only active when combined with the protein molecule of the enzyme, apoenzyme, to form a holoenzyme. Such prosthetic groups are known as coenzymes

$$\text{apoenzyme} + \text{coenzyme} \rightleftharpoons \text{holoenzyme}$$

Many vitamins function as coenzymes, helping to catalyze biochemical reactions vital to growth and structural maintenance.

NUCLEOPROTEINS

These are an extremely important class of conjugated proteins in which the prosthetic groups are long chain polymers called *nucleic acids*. The chromosomes of cell nuclei are nucleoproteins. The genes that control hereditary characteristics are sections of the long chain polymers. The nucleic acids thus carry the genetic code. They ensure the production of the particular proteins which characterize each cell for all the different types of organism; they pass on the genetic information when a cell divides; they also transmit the blueprint of hereditary characteristics from one generation to the next.

ROLE OF NUCLEIC ACIDS IN HEREDITY

The master plan for constructing the correct set of proteins for an organism is contained in the nucleic acids, DNA, *d*eoxyribo*n*ucleic *a*cid, and RNA, ribo*n*ucleic *a*cid. These polymers have a sugar-phosphate backbone.

---Sugar—phosphate—sugar—phosphate—sugar—phosphate---

The sugar is deoxyribose in DNA and ribose in RNA. Attached to each sugar residue is a molecule of a base, so that the polymer chain can be represented like this

```
    base               base                base
     |                  |                   |
---Sugar—phosphate—sugar—phosphate—sugar—phosphate---
```

The chain is coiled in the form of a right-handed helix. Each nucleic acid employs only four different bases. DNA uses adenine, A, cytosine, C, guanine, G, and thymine, T. RNA uses uracil, U instead of thymine. The genetic code is simply the sequence of these bases along the polymer chain. The sequence of bases determines what proteins will be made and hence what characteristics the organism will have. We will see in a moment how this code operates.

Watson and Crick found that in DNA two helices are intertwined in a special way. Adenine in one chain is always hydrogen-bonded to thymine in the other chain; similarly cytosine is always paired with guanine. The bases always show the same complementary pairing. The structure of this double helix is shown schematically in Fig. 25.3.

The double helix offers a simple explanation for cell replication and the transfer of the genetic code. Before mitosis (cell division) each DNA double helix in the nucleus separates into its two halves and as it does

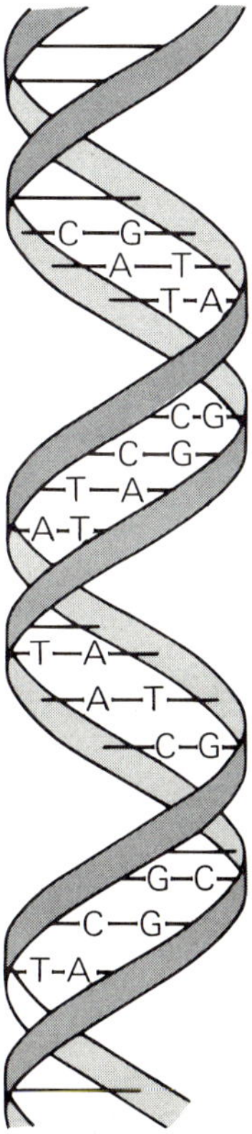

Figure 25.3
DNA showing double helix joined by paired bases.

so the bases on each single strand join up with fresh molecules of the complementary bases (that is, A with T and C with G) which are freely available in the nucleus. The sugar-phosphate backbone is simultaneously added so that a new polymer chain is coiled around each helix, identical with the one that has been removed. In this way, each double helix duplicates itself so that, when the cell divides, the two daughter cells will each have the same set of DNA double helices, and hence the same genetic code as the parent cell. They will, therefore, be replicas of it.

How does the code operate? Or, putting the question another way, how are the various amino acids brought together in the right order to make a particular protein? The answer lies in the sequence of the four bases adenine, A, cytosine, C, guanine, G, and uracil, U. in RNA. The molecules of RNA are single helices and are made in the nucleus from DNA templates by much the same process that occurs when DNA duplicates itself. It is now known that three consecutive bases, called a codon form a code for a particular amino acid. Thus CAU is the code for histidine. Some amino acids have more than one codon. Examples are given in Table VIII.

Suppose a section of RNA has the following sequence of codons:

G C A G A A C A C A A A U U U

This then ensures the following sequence of amino acids in the synthesized protein:

Ala	Glu	His	Lys	Phe
G C A	G A A	C A C	A A A	U U U

The majority of the RNA molecules pass out of the nucleus into the cytoplasm (messenger RNA) where they carry out most of their work of protein building in the ribosomes. DNA stays at home in the nucleus doling out copies of its master plan to the RNA labor-force.

Clearly, any change or mutation in the nucleic acid structure, that is in the genes, could lead to a change in protein structure, which in turn could produce significant changes in the organism. Some babies have a nucleic acid defect which prevents the formation of the enzyme needed to convert the amino acid, phenylalanine, into tyrosine. The absence of this enzyme leads to the overproduction of phenyl pyruvic acid which causes mental retardation in the infant. When the baby is 6 to 8 weeks old, the phenyl pyruvic acid can be detected in the urine, a condition known as phenylketonuria or PKU. Earlier diagnosis is clearly preferable

Table VIII. AMINO ACID CODONS

AMINO ACID	CODONS
alanine, Ala	GCA GCC GCU GGG
glutamic acid, Glu	GAA GAG
histidine, His	CAC CAU
lysine, Lys	AAA AAG
phenylalanine, Phe	UUU UUC

and is possible by testing for excessive phenylalanine in a blood sample taken a few days after birth. The brain damage, once caused, cannot be repaired but it can be prevented by excluding as much phenylalanine from the protein diet as possible. This means no milk, eggs, bread, or meat. Clearly, unless special supplements are used, the treatment can be as dangerous as the condition it is intended to prevent.

ROLE OF NUCLEIC ACIDS IN VIRUSES

Before leaving the topic of nucleic acids, brief mention should be made of viruses and the way they act. They consist almost entirely of nucleic acids in a protein jacket. Some are DNA viruses and others are RNA. They are virulent because their nucleic acids take over the host cells which they invade. They use the cells' resources to make their own proteins and nucleic acids, so the virus multiplies and the cells die.

26

Enzymes

Enzymes are concerned in the great majority of chemical reactions which occur in the body, including those involved in digestion. The word enzyme literally means in yeast (Gr. *en*, in, +Gr. *zymē*, yeast). The ability of yeast to convert sugar into alcohol depends on the presence of an enzyme called zymase which, like most enzymes, is a protein.

Enzymes are catalysts. They facilitate chemical reactions without themselves being used up in the process. Catalysts in inorganic chemistry are not specifically adapted to assist in a particular reaction. Platinum, for example, acts as a catalyst in quite a wide range of chemical reactions. Enzymes on the other hand, which may be considered as organic catalysts, have a structure which is specifically adapted to either a single reaction or to a very limited range of almost identical reactions.

This specificity is a characteristic feature of enzymes. All enzymes will, however, speed up a reaction in either direction. For example, the enzyme *carbonic anhydrase* facilitates the combination of carbon dioxide and water to form carbonic acid.

$$CO_2 + H_2O \xrightarrow[\text{anhydrase}]{\text{carbonic}} H_2CO_3$$

This is the reaction which occurs when carbon dioxide, produced as a result of tissue respiration, diffuses from the tissue cells through the plasma into the red cells. In the lungs the same enzyme, carbonic anhydrase facilitates the decomposition of carbonic acid into carbon dioxide and water.

$$CO_2 + H_2O \xleftarrow[\text{}]{\text{carbonic anhydrase}} H_2CO_3$$

The reversible reaction, typical of most reactions in the body, is

$$CO_2 + H_2O \rightleftharpoons H_2CO_3$$

(remember that a double arrow is used instead of an equal sign to show reversibility).

The chemical reaction actually proceeds in both directions at the same time. But if the carbon dioxide is removed by the lungs, then practically none remains to participate in the reaction so that the direction of the reaction is almost entirely from right to left. In the tissues the reverse applies. The direction of the reaction is thus controlled by the concentration of the reacting substances.

Enzymes are often named after the substance on which they act (the substrate) with "-ase" added. Thus proteinase is an enzyme which splits protein. Lipase (Gk. *lipos*, fat) is an enzyme which splits fat, amylase (L. *amylum*, starch) is a starch-splitting enzyme.

Sometimes enzymes are named after the reaction that they promote. Thus an oxidase promotes oxidation and a deaminase promotes removal of an amino or $-NH_2$ group. A number of enzymes, particularly those concerned with digestion, were given their names before the present terminology was developed. So the starch-splitting enzyme in saliva originally called ptyalin is now called *salivary amylase*. The protein-splitting enzymes in stomach secretion and pancreatic juice originally called pepsin and trypsin are now called *gastric proteinase* and *pancreatic proteinase*. We note that the old names are still used by many people. The old names for the other pancreatic enzymes amylopsin, or starch-splitting, and steapsin, or fat-splitting, have now been dropped in favor of *pancreatic amylase* and *pancreatic lipase*. The activity of enzymes is greatly affected by the pH or acidity, the electrolyte or salt concentration, the temperature, toxic substances and other factors.

pH, HYDROGEN ION CONCENTRATION

Some enzymes will only function in an acid medium. Gastric proteinase (pepsin) is inactivated by alkali and functions best at about pH 1.5, which is a remarkable degree of acidity. In contrast, pancreatic amylase is inactivated by strong acids; it functions best in a neutral solution, pH 7.0, and tolerates an alkaline medium.

ELECTROLYTE CONCENTRATION

Enzyme activity is greatly affected by the concentration of certain ions. Amylases will only function in the presence of chlorine ions, $Cl^{\ominus}$. Pancreatic proteinase (trypsin) is activated by calcium ions, $Ca^{2\oplus}$. The action of phosphatase, the enzyme involved in the calcification of bone, is accelerated by magnesium ions, $Mg^{2\oplus}$. Ions and other substances which promote enzyme activity are known as coenzymes. The term coenzyme is also used for a prosthetic group which must be attached to an enzyme protein molecule before it can act as a biological catalyst.

TEMPERATURE

Most enzymes in the human body function best at a temperature of about 37°C. Lowering the temperature slows the reaction. Slightly raising the temperature speeds up some reactions. A greater rise in temperature slows them. Temperatures of 60°C or more permanently destroys the activity of human enzymes (Fig. 26.1). The enzymes of certain bacteria (thermophilic bacteria) can function well at high temperatures whereas those of fungi function best at rather low temperatures of about 20°C.

TOXIC SUBSTANCES

Enzymes are inactivated by poisons such as cyanides which can cause rapid fatality due to their inactivation of tissue respiration enzymes. Formaldehyde, alcohol, acetone and heavy metal salts such as mercury(II) chloride are all used as fixatives to preserve tissue from decay. They act by denaturing the enzymes, as in Fig. 26.1, and so destroying their activ-

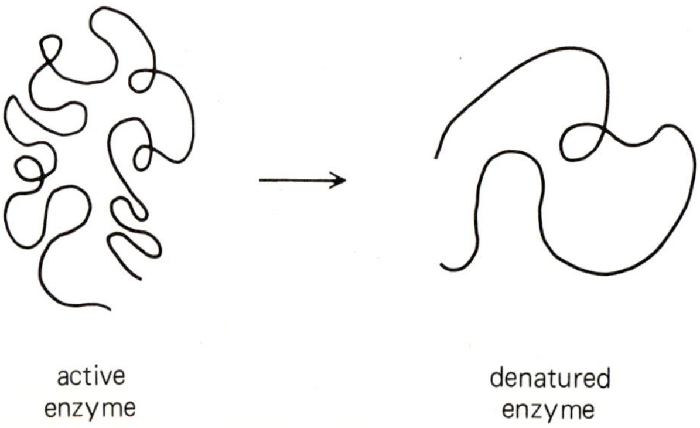

Figure 26.1
Diagram of denaturation (irreversible inactivation) of an enzyme.

ity. The halogens, that is fluorine, bromine, chlorine, and iodine, are also toxic for certain enzymes.

OTHER FACTORS

The initial speed of an enzyme-catalyst reaction is directly proportional to the enzyme concentration. The speed of the reaction is also increased by raising the concentration of the substrate (substance on which the enzyme acts), but only up to the point where the enzyme is fully saturated with substrate. Beyond that point, increasing the concentration of the substrate slows the reaction.

The shape of the enzyme molecule plays an important part in its interaction with the substrate. There is evidence that the substrate molecule actually fits into the configuration of the enzyme molecule rather like a lock and key (Fig. 26.2). Another substance which has a sufficiently similar shape to the substrate to enable it to fit into the enzyme molecule could block an active site so that the enzyme molecule could no longer react with the substrate (Fig. 26.3). This is known as compe-

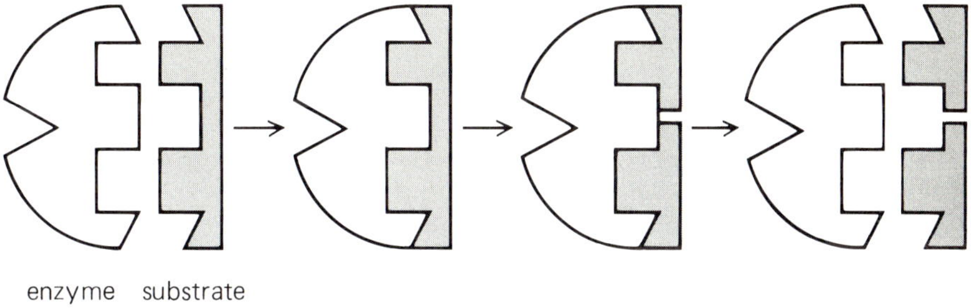

Figure 26.2
Schematic of substrate-splitting enzyme with shape to fit substrate molecule.

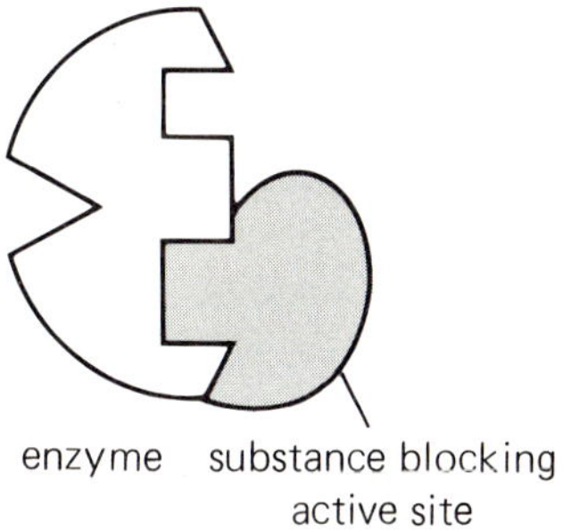

Figure 26.3
Schematic of competition inhibition: substance blocks active site on enzyme.

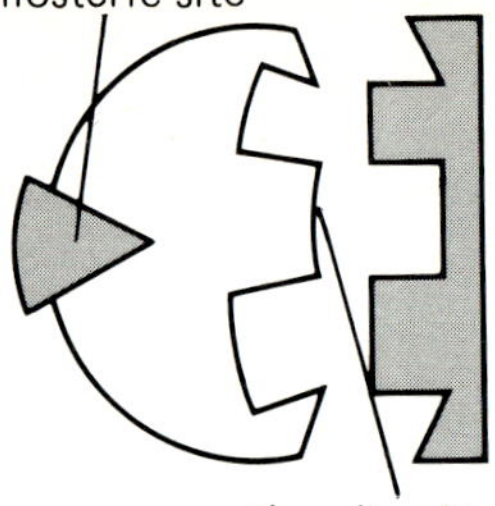

Figure 26.4
Schematic of allosteric inhibition: substance combines with enzyme away from active site altering shape of enzyme molecule.

titon-inhibition. Enzyme activity can also be inhibited by a substance combining with the enzyme molecule at a site away from the active site in such a way as to alter the shape of the enzyme to a degree that prevents the substrate from reaching an active site. This is known as allosteric inhibition (Fig. 26.4). In the body it provides a means of controlling enzyme activity, for example, in preventing excessive production of an amino-acid such as isoleucine.

CLASSES OF ENZYMES

Enzymes may be classified according to the types of reaction which they catalyze. They may be conveniently considered under the following four headings:

1. *Hydrolyzing enzymes:* This group includes all the enzymes concerned in digestion (Chapter 27).
2. *Oxidation-reduction enzymes:*
   oxidases—promote oxidation in the presence of free oxygen.
   dehydrogenases—remove hydrogen from substrate and transfer it to a hydrogen acceptor. These are involved in the Krebs cycle.
   peroxidases—remove oxygen from hydrogen peroxide H_2O_2 or organic peroxide and transfer it to a substrate which acts as an oxygen acceptor.
   catalase—catalyzes the breakdown of hydrogen peroxide into water and molecular oxygen

$$2H_2O_2 \xrightarrow{\text{catalase}} 2H_2O + O_2$$

in all aerobic (or oxygen using) cells and removes hydrogen peroxide that otherwise would be toxic for protoplasm.

3. *Enzymes that separate two carbon atoms*
   Examples are the carboxylases of cell respiration, for example, oxaloacetic carboxylase splits CO_2 off 4-carbon atom oxaloacetic acid to form 3-carbon atom pyruvic acid, and pyruvic carboxylase splits CO_2 off 3-carbon atom pyruvic acid to form 2-carbon atom acetaldehyde.
4. *Miscellaneous enzymes*
   These are enzymes which do not fit into the above three groups. Examples are the isomerases, one of which converts glucose-6-phosphate into fructose-6-phosphate, an essential step in the respiratory breakdown of glucose. Another member of the miscellaneous group is the carbonic anhydrase quoted at the beginning of this chapter, which assists the removal of carbon dioxide from the tissues and from the lungs.

27

Digestion

The essence of digestion is the chemical breakdown of food by enzymes. In order to function effectively the enzymes require the right chemical conditions, for example, gastric acidity and small intestinal alkalinity. There are also a number of important physical factors. These include involuntary muscular contractions, as in peristalsis, partially involuntary muscular contractions as in swallowing (deglutition), and voluntary muscular contractions as in biting and chewing (mastication). An important underlying factor is the anatomy of the alimentary tract (Fig. 27.1).

The alimentary tract consists of a long tube in which certain parts are dilated, notably the stomach and to a lesser degree the mouth and the large intestine, and certain sites are constricted. The constrictions are produced by valves consisting mainly of circular muscles, called the sphincters (Gr. *sphigkter*, shut tight). The sphincter between the stomach and duodenum is called the pylorus (Gr. *pyloros*, gatekeeper); between the small and large intestine is the ileo-caecal valve and guarding the lower end of the gut is the anal sphincter. The sphincters relax to allow the contents to pass through at appropriate stages of digestion.

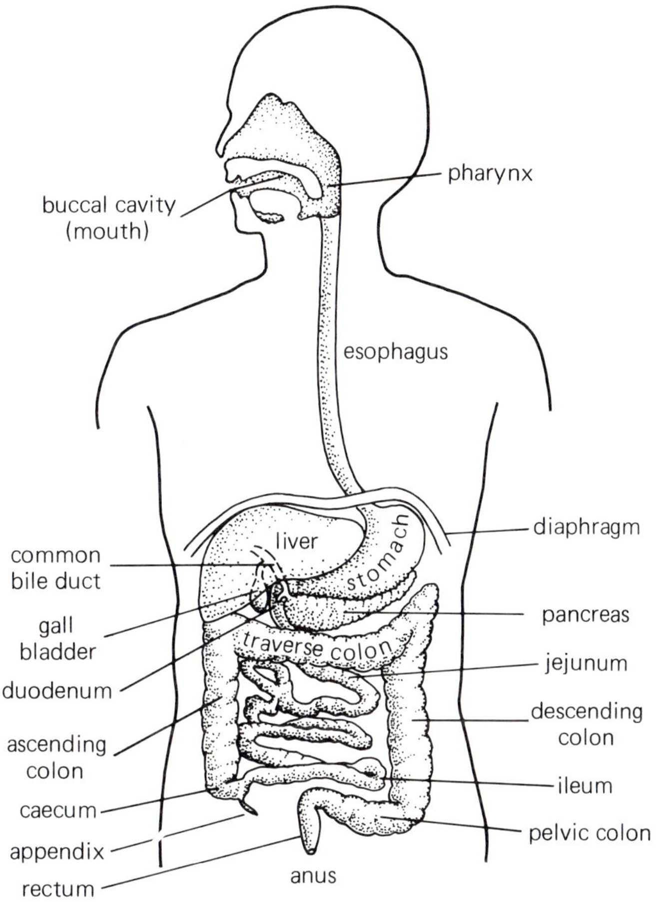

Figure 27.1
Alimentary tract.

DIGESTION IN THE MOUTH

Mastication is the first step in the digestive process, physically breaking the food material down into small fragments and enabling it to be impregnated with saliva. The main enzyme in saliva is an amylase (starch-splitter) known until recently as ptyalin. Given sufficient time to act this enzyme can break down the large molecules of starch through a series of intermediary molecules of successively smaller size to produce the disaccharide (12 carbon atom sugar) maltose. Normally, however, it does not have sufficient time to produce its full effect as it is inactivated by the strong acid in the stomach. When eating a dinner, much of the carbohydrate is consumed at the end of the meal as the sweet. Being the last to enter the stomach it may remain for a while near the cardia, thus escaping acid inactivation of the amylase for half an hour or so and allowing at least partial digestion of starch to occur. The steps by which amylase breaks down starch can be studied by collecting saliva and adding it to a suspension of boiled starch in a test tube. The reason for boiling the starch is that salivary amylase has no effect on cellulose. So the starch has to be cooked first to rupture the cellulose envelope of the starch grains.

The starch suspension is opalescent and gives a blue color with iodine. After salivary amylase has acted on the starch for a short while it loses its opalescence, indicating that the starch has been converted into soluble form although still giving a blue color with iodine. The soluble starch is then broken down into a dextrin (the slightly sweet material on the surface of toast) which gives a red color with iodine and so is called *erythro*dextrin (Gr. *erythros*, red). Further amylase action produces a dextrin with smaller molecules which gives no color reaction with iodine, *achromo*dextrin or *achroo*dextrin (Gr. *a*, without, + Gr. *chromos*, color). This is finally converted into maltose and isomaltose. Saliva also contains a small amount of maltase which converts maltose to glucose. The steps by which starch is broken down if salivary digestion proceeds to completion may be represented as follows:

boiled starch
↓
soluble starch
↓
erythrodextrin + maltose
↓
achroödextrin + maltose

} HYDROLYSIS BY AMYLASE

↓
maltose + isomaltose
↓
glucose

} HYDROLYSIS BY MALTASE

It will be remembered that hydrolysis involves the splitting of the sugar molecule and the addition of a water molecule, namely

$$\underset{\text{MALTOSE}}{C_{12}H_{22}O_{11}} + H_2O \xrightarrow{\text{maltase}} \underset{\text{GLUCOSE}}{2C_6H_{12}O_6}$$

Often, however, the salivary amylase has too little time for much of the above to take place. So one of the main functions of saliva in the digestive process is the lubrication of the food bolus to facilitate its passage through the esophagus.

DIGESTION IN THE STOMACH

The stomach is an organ capable of very considerable expansion (Fig. 27.2) and serves as a temporary storage place for food immediately after it is consumed. At the same time the food is kneaded by waves of muscular contraction of the stomach wall and impregnated with gastric juice. This is secreted by the numerous glands in the gastric mucosa and contains hydrochloric acid, mucin and pepsin.

The hydrochloric acid is a strong mineral acid with a pH of 0.9 to 1.5 as secreted by the parietal cells in the stomach wall (Fig. 27.3). It becomes diluted to some degree, however, by the stomach contents. In some way not yet fully understood these cells manage to produce this very strong acid from the chloride ions, $Cl^{\ominus}$ in the blood without themselves suffering corrosion. The acid is formed in small canals within the cells, intracellular canaliculi, which connect up with extracellular canaliculi; through these channels the acid flows into the lumen of the gland (Fig. 27.3 (a)) and thence out into the stomach cavity.

Gastric juice

The gastric juice contains a number of enzymes that are secreted by the chief cells in the gastric glands (Fig. 27.3(b)). The main enzyme is gastric

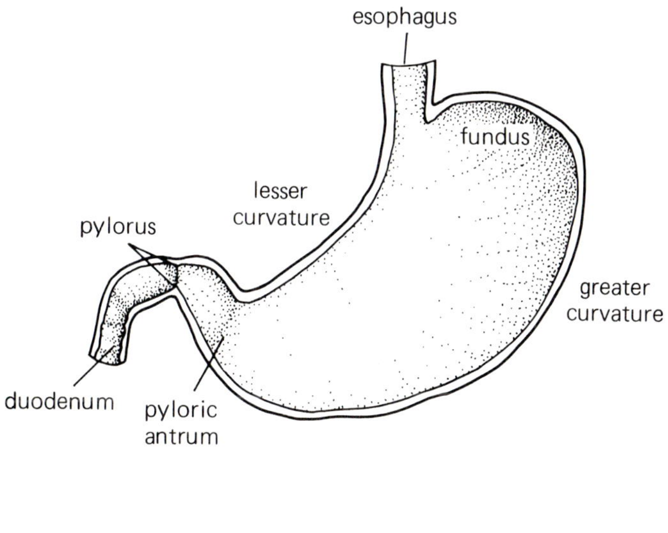

Figure 27.2
Stomach.

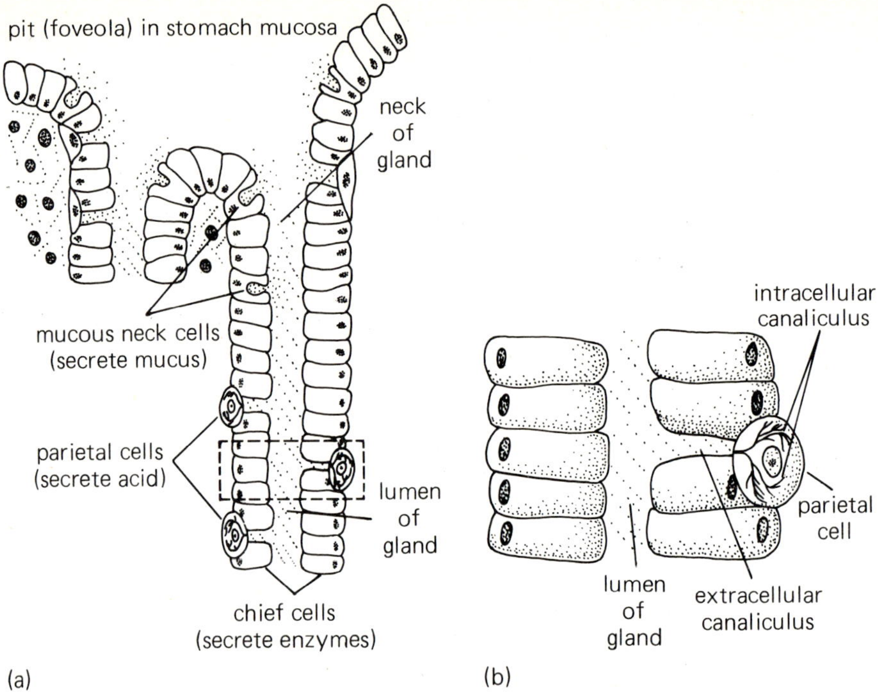

Figure 27.3
(a) gastric glands; (b) detail of gastric gland.

proteinase or pepsin which splits proteins into smaller molecules of proteoses and peptones. Its activity is greatest at pH 1.5 so that it is ideally suited for gastric digestion. Another enzyme is rennin which curdles milk, splitting it into curds and whey. There is also a gastric lipase which in the adult has a very weak fat-splitting effect, acting only on finely emulsified fat such as cream. It is possible that in infants its activity is of greater importance.

In the neck of the gastric glands are mucous glands (Fig. 27.3(a)). The mucus which they secrete in considerable quantity tends to cling to the gastric mucosa, protecting it from physical damage by hard, irregular portions of food and from chemical damage by the gastric acid and normally preventing peptic ulcers from developing.

The main stimulus to secretion of gastric juice is the entry of food into the stomach. Psychic stimuli such as the sight, smell, taste or thought of food are also effective. A third type of stimulus arises from the breakdown products of food during intestinal digestion.

The amount of acid secreted in the stomach varies to some degree in normal people. In people with duodenal ulcer the acid content tends to be higher than normal, hyperchlorhydria. In carcinoma of stomach it is lower than normal and may be absent, achlorhydria. In pernicious anemia it is characteristically absent.

The acid is measured by means of a test meal. The fasting subject swallows a Ryle's stomach tube and a mark on the tube shows when it has passed far enough. The fasting juice is aspirated with a 20 cm^3 syringe. Gastric secretion is then stimulated by the injection of pentagastrin (Peptavlon, I.C.I.) subcutaenously. This is a powerful gastric stimulant which has now virtually replaced gruel, toast, tea, alcohol or histamine injections in the standard test meal. Samples of gastric contents are then withdrawn every 15 minutes for chemical analysis (Fig. 27.4).

The gastric acidity is expressed as free HCl, which is hydrochloric acid alone, and total acidity, which is HCl plus other acids present. In pernicious anemia there is typically no free acid present even after the

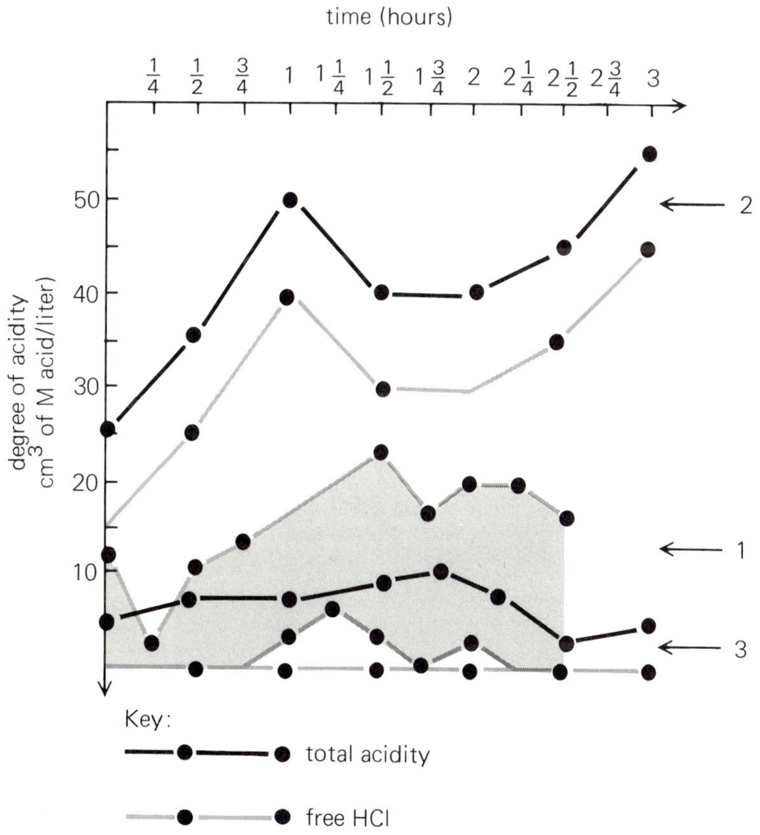

Figure 27.4

Fractional test meal: (1) normal (shaded area); (2) increased hydrochloric acid (hyperchlorhydria); (3) no hydrochloric acid (achlorhydria). (Modified from *Special Tests and Their Meanings*, 9th edition, by D. M. D. Evans. Courtesy Faber & Faber, London.)

most powerful histamine stimulus. This failure to produce acid reflects the wasted (atrophic) state of the gastric mucosa which is in fact responsible for the anemia. The atrophic stomach fails to secrete a substance called intrinsic factor which in normal people combines with vitamin B_{12}, enabling it to be absorbed in the small intestine.

DIGESTION IN THE INTESTINE

After the preliminary digestion in the stomach the pyloric valve opens and the contents pass on into the small intestine where the major part of digestion and absorption occurs. The first twelve inches of the small intestine is called the duodenum (Fig. 27.5) into which open the bile duct and the main pancreatic duct which join at a common orifice, the ampulla of Vater.

Bile

Bile is secreted by the liver. Its color is due to the presence of the bile pigments, orange bilirubin and green biliverdin. These are the iron-free breakdown products of hemoglobin. In the gut they oxidize to brown stercobilin, giving the intestinal contents and feces their characteristic color. In addition to these and other breakdown products there are the bile salts, sodium glycocholate and taurocholate. Bile salts greatly assist fat digestion and absorption. They act by emulsifying fat, breaking it up into tiny droplets. They also activate pancreatic lipase which splits

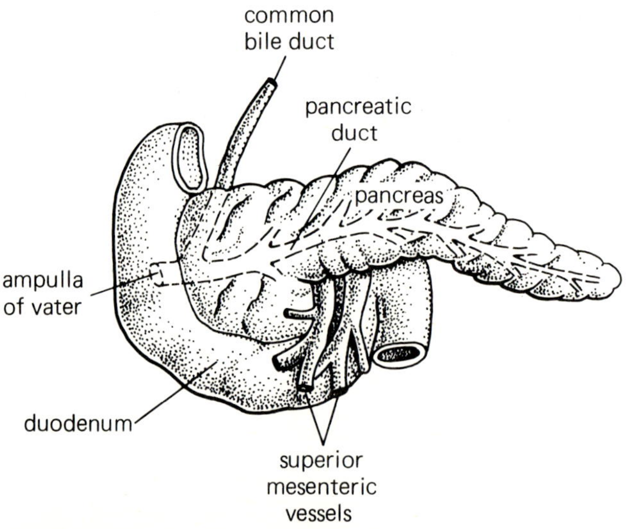

Figure 27.5
Duodenum and pancreas.

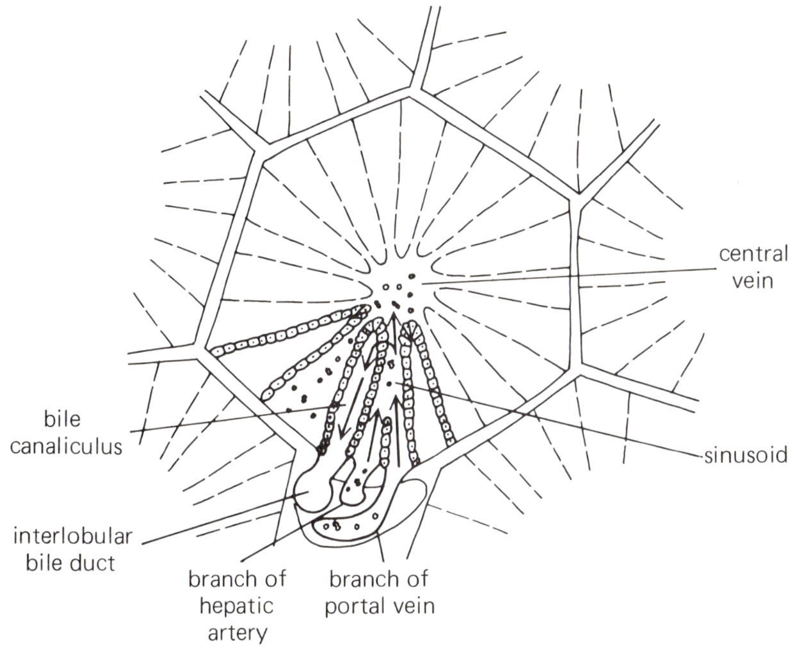

Figure 27.6
Liver lobule.

fat into fatty acids and glycerol or glycerine. They then combine with the fatty acids, rendering them soluble and enabling them to be absorbed. Their effect on fats is in some ways similar to that of detergents.

Bile salts are synthesized in the liver cells from *cholesterol*. This is converted into *cholic* acid, which has a similar structure, and combined with the amino-acids *glycine* and *taurine* to form *glycocholic* and *taurocholic* acids. The bile salts are the sodium salts of these acids. This is just one example of the fantastic variety of chemical operations that the liver can undertake. It can be seen from Fig. 27.6 that the liver cells are well supplied with blood, from which they obtain the raw materials, and can also secrete bile directly into small channels, the bile canaliculi, which open into larger channels, the bile ducts, which eventually open into the common bile duct and thence, through the ampulla of Vater, into the duodenum (Fig. 27.5).

The secretion of bile is disturbed in the various forms of jaundice (see Chapter 30, Laboratory examination of urine, bile pigments and salts).

Pancreatic juice

The pancreatic juice is secreted by the pancreas (Fig. 27.5), known in butcher's shops as sweetbread. The formation of pancreatic juice by the cells will be described later (Chapter 28). It is alkaline due to the presence of sodium carbonate and bicarbonate. It contains a number of enzymes, the main ones being lipase (or steapsin), amylase (or amylopsin) and protease or proteinase (or trypsin). The names in parentheses are the old terminology and with the exception of trypsin are falling into disuse.

The splitting of fats into fatty acids and glycerol by lipase, assisted by the action of bile salts, has already been discussed. The steps by which pancreatic amylase splits starch into maltose are the same as those described for salivary amylase, but it is a more powerful enzyme and has a longer time in which to act. So most of the digestion of starch takes place in the small intestine. Some maltase is also present in pancreatic juice converting part of the maltose into glucose.

The pancreatic protein-splitting enzyme is secreted in an inactive form known as trypsinogen. It only becomes active after coming into contact with intestinal juice (succus entericus) which contains an activator (enterokinase).

$$\underset{\text{INACTIVE PRECURSOR}}{\text{trypsinogen}} + \underset{\text{ACTIVATOR}}{\text{enterokinase}} \longrightarrow \underset{\text{ACTIVE PROTEASE}}{\text{trypsin}}$$

Trypsin acts both on the breakdown products of protein produced by gastric digestion and on any unsplit protein which may have escaped digestion by pepsin in the stomach. Unlike pepsin that functions only in an acid medium, trypsin acts only in an alkaline medium with an optimum pH of 8.5. Like pepsin it can convert proteins into proteoses and peptones, but it can also continue the breakdown process to a further stage with the production of polypeptides (mentioned in Chapter 24). Trypsin digestion can progress to the stage of tetrapeptides, tripeptides, and dipeptides which contain four, three, and two amino-acid residues, respectively.

Intestinal juice

Intestinal juice or succus entericus (L. *succus*, juice + L. *entericus*, intestinal) is secreted by cells in little pits situated in the wall of the small intestine (the crypts of Lieberkühn, Fig. 27.7). It contains mucus, secreted by goblet cells in the neck of the crypts; the alkaline salts sodium carbonate and bicarbonate, presumably secreted by cells halfway down the crypts; and enzymes, believed to be secreted by "Paneth cells" at the base of the crypts. The latter cells contain granules like those seen in the pancreatic cells.

In addition to enterokinase, which as we have said activates trypsinogen, there are enzymes which can complete the final stages of digestion of fat, carbohydrate and protein. The fat-splitting enzyme lipase is

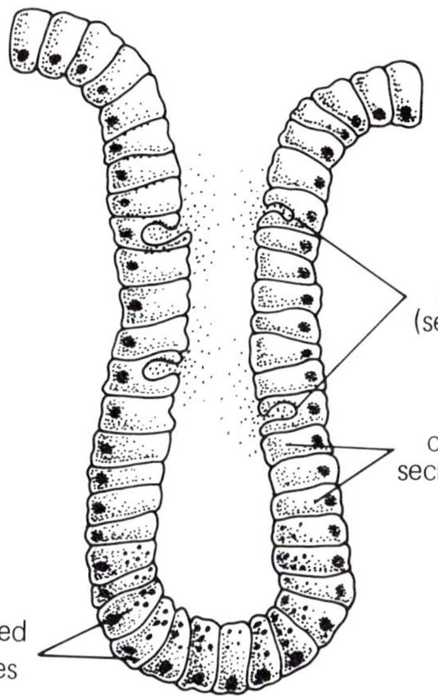

Figure 27.7
Crypt of Lieberkühn.

not so powerful as that from the pancreas but is still capable of splitting 70% of the fat in the diet without assistance from pancreatic lipase. There are three carbohydrate-splitting enzymes called sucrase, maltase and lactase. Sucrase acts on cane sugar, sucrose:

$$\underset{\text{SUCROSE}}{C_{12}H_{22}O_{11}} + H_2O \xrightarrow{\text{sucrase}} \underset{\text{GLUCOSE}}{C_6H_{12}O_6} + \underset{\text{FRUCTOSE}}{C_6H_{12}O_6}$$

Maltase acts on beet sugar, maltose,

$$\underset{\text{MALTOSE}}{C_{12}H_{22}O_{11}} + H_2O \xrightarrow{\text{maltase}} \underset{\text{GLUCOSE}}{2C_6H_{12}O_6}$$

Lactase acts on maple sugar, lactose:

$$\underset{\text{LACTOSE}}{C_{12}H_{22}O_{11}} + H_2O \xrightarrow{\text{lactase}} \underset{\text{GLUCOSE}}{C_6H_{12}O_6} + \underset{\text{GALACTOSE}}{C_6H_{12}O_6}$$

In general terms, each of these three enzymes acts on disaccharides (12 C atoms) converting them to monosaccharides (6 C atoms), a form in which they can readily be absorbed.

The enzymes which are responsible for the final stages of protein digestion are the peptidases, previously called erepsin (Gr. *ereptomai*, to feed upon). These enzymes have no action on unsplit protein. It is only after trypsin has broken the protein down to the peptide stage that peptidases become effective. They complete the protein breakdown by splitting the peptides into their constituent amino-acids which are then absorbed. Intestinal juice also contains enzymes which digest nucleic acid and its derivatives; each enzyme is named after the substance which it digests, namely, nuclease, nucleotidase and nucleosidase, which, respectively, digest nucleic acids, nucleotides and nucleosides.

TYPES OF DIGESTION

Fat digestion

Fat is split by lipases into fatty acids and glycerine (or glycerol). The most powerful lipase is that produced by the pancreas. Its action is assisted by the bile salts which emulsify the fats, activate the lipase and promote fatty acid absorption. The lipase of intestinal juice can also digest most of the fat in a normal diet. Gastric lipase is relatively weak and is only able to act on fats already in a state of fine emulsion before ingestion, as in milk.

Carbohydrate digestion

Starch is mainly digested by pancreatic amylase which hydrolyzes it through the dextrins as far as the disaccharide maltose. Salivary amylase is potentially similar but is normally inactivated by gastric acid before digestion has progressed very far. The breakdown of disaccharides to monosaccharides is mainly by the maltase, sucrase and lactase in the intestinal juice.

Protein digestion

The acid gastric juice with its proteinase (pepsin) splits proteins into proteoses and peptones. Pancreatic proteinase is secreted as inactive trypsinogen which is converted into active trypsin by the enterokinase of intestinal juice. Trypsin can act on either intact protein or the proteoses and peptones resulting from pepsin digestion, breaking them down to the peptide stage. The peptides are split by the peptidases in intestinal juice into their constituent amino-acids ready for absorption. Intestinal juices also contain nuclease and other enzymes which digest nucleic acid and its derivatives.

28

Cell Chemistry

Just as atoms all have a similar pattern with a dense central nucleus surrounded by a space containing orbiting electrons, so also we find a general similarity of pattern in all living cells. Typical animal and vegetable cells have a dense central nucleus, very much more complex than that of the atom, surrounded by a space containing semiliquid cytoplasm which is enclosed in a cell membrane (Fig. 28.1).

Each cell is capable of undertaking a fantastic variety of chemical reactions, made possible by a vast number of different enzymes. To prevent unwanted interference between one chemical reaction and another the cell cytoplasm is divided up into a number of different compartments or organelles. One example is the mitochondrion (noted by M in Fig. 28.1), the site of cellular respiration. This contains a series of enzymes that progressively oxidize glucose into CO_2 and water. The cell is dependent on the mitochondria for its supply of energy. So the more active a cell, the more mitochondria it requires.

A second type of organelle is the lysosome (noted by L in Fig. 28.1). This contains potent enzymes, capable of digesting the cell itself; it has been compared to a suicide capsule. However, in a normal cell the ac-

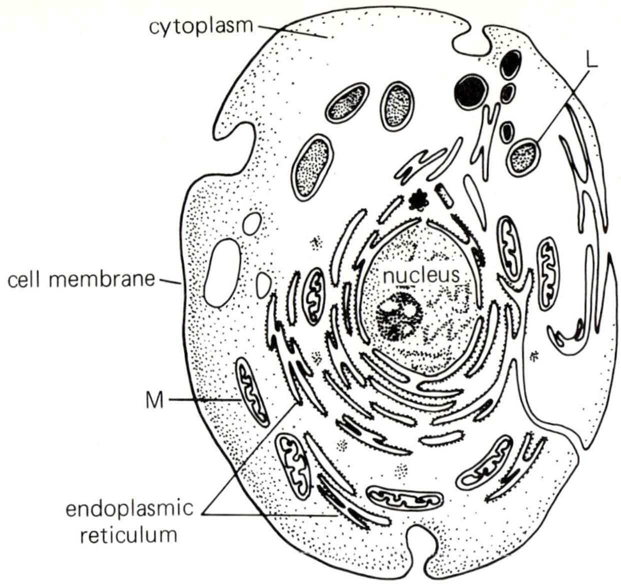

Figure 28.1
Mammalian cell (ultrastructure): M-mitochondria; L-lysosomes.

tion of these enzymes is kept in check by their being enclosed in a membrane from which they cannot escape. Their activity is harnessed to the process of cellular digestion. The first step in this process is the formation of a food vacuole. This is produced by indentation of the cell surface at a site where fluid food material is present. Enlargement of the indentation progresses until the fluid material is completely enclosed by the indented membrane (pinocytosis). The food vacuole so produced then coalesces with a lysosome to produce a digestive vacuole in which the lysosomal enzymes interact with the food material. Solid food material may similarly be ingested (phagocytosis) to form a food vacuole and the contents subjected to lysosomal digestion (Fig. 28.2).

Another important cell organelle system is that involved in secretory activity. The exact details vary with the type of secretion produced. In the cells forming the pancreatic juice, for example, several stages have been recognized (Fig. 28.3). The cytoplasm of the cell contains spaces, cisternae, into which proteins are secreted by the surrounding granular organelles, ribosomes. The spaces are continuous with a network of channels, the endoplasmic reticulum, part of which has long been known as the Golgi apparatus. The secretion is then stored in globules, the zymogen granules, until needed. Finally the contents of the granules are ejected out through the cell membrane into the gland lumen as the pancreatic juice.

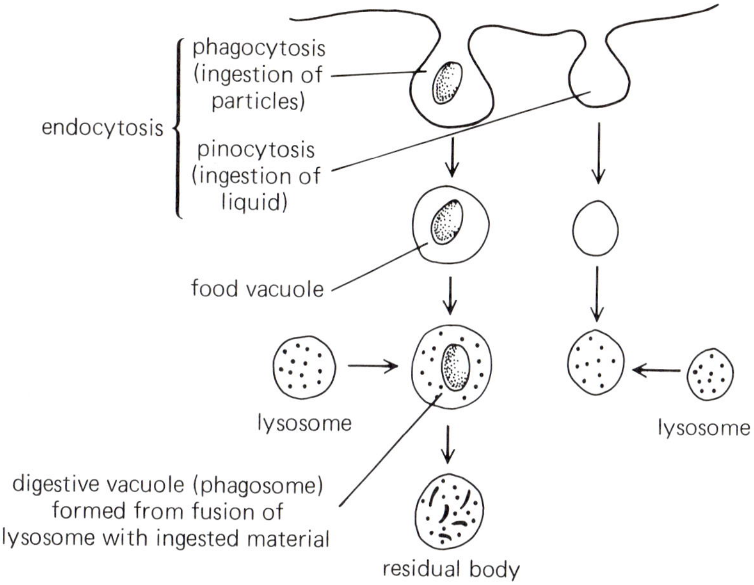

Figure 28.2
Lysosomal digestion. The digestive vacuole (phagosome) is formed from fusion of a lysosome with a food vacuole. In protozoa the contents of the residual body have been seen to be ejected—cellular excretion or "defecation"—but this has not been observed in mammalian cells.

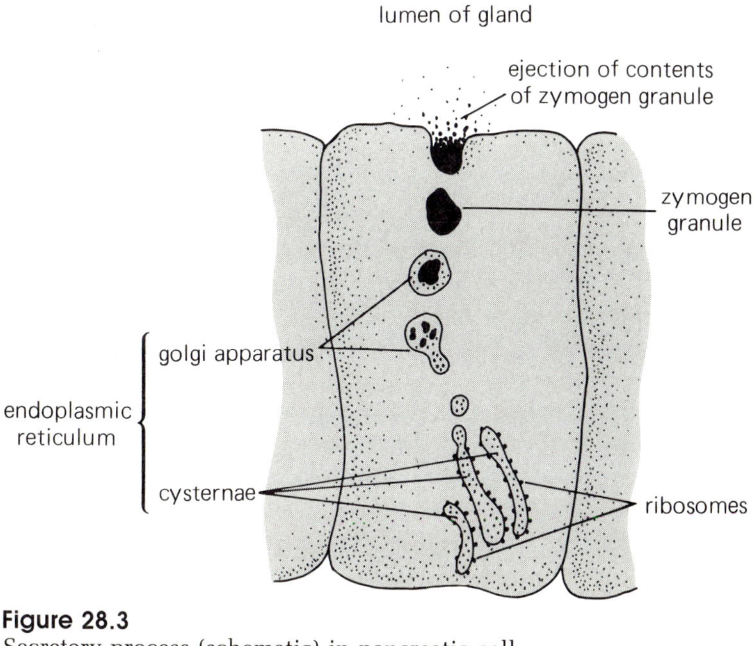

Figure 28.3
Secretory process (schematic) in pancreatic cell.

CELL METABOLISM

Cell metabolism is the sum of all the chemical reactions taking place within the cell. It includes the process of catabolism by which substances are broken down, often with the release of energy as in cell respiration, and of *anabolism* by which new substances are formed, as in the synthesis of the numerous different proteins, a process involving the expenditure of energy.

Cell Respiration

The purpose of respiration is to provide energy so that the body can function. The energy is produced by the oxidation of carbohydrates, fats, and proteins obtained from the food we eat. Thus, the oxidation of one mole (180 g) of glucose, $C_6H_{12}O_6$ involves the release of 2850 kilojoules of energy.

$$C_6H_{12}O_6 \quad 6O_2 \longrightarrow 6H_2O + 6CO_2 + 2850\text{kJ}$$

Inside the cell the release of this enormous amount of energy all at once with the heat and flame involved would be catastrophic. So instead the energy is released in a series of small steps within the mitochondria (Figure 8.5). The energy released at each step is taken up by molecules of adenosine diphosphate, ADP which in the process are converted to molecules of adenosine triphosphate, ATP.

$$-\text{(P)} + \underset{\text{ADP}}{\text{adenosine}-\text{(P)}-\text{(P)}} + \underset{\text{ENERGY}}{33.5\text{ kJ}} \longrightarrow \underset{\text{ATP}}{\text{adenosine}-\text{(P)}-\text{(P)}-\text{(P)}}$$

Energy can be released when required by reversing the above reaction.

$$\underset{\text{ATP}}{\text{adenosine}-\text{(P)}-\text{(P)}-\text{(P)}} \longrightarrow -\text{(P)} + \underset{\text{ADP}}{\text{adenosine}-\text{(P)}-\text{(P)}} + \underset{\text{ENERGY}}{33.5\text{ kJ}}$$

The ATP may be considered as the cell's energy currency. It is spent on energetically expensive reactions such as those involved in protein synthesis and muscular contraction. If the cell becomes bankrupt of ATP then its functions become disturbed and may cease. For example, the brain cells can no longer function in severe hypoglycemia (low blood sugar) and so unconsciousness results. If glucose is injected intravenously energy currency is restored and consciousness returns, provided the delay is not long enough for irreversible changes to occur. The critical role of oxygen for cell energy and vitality has already been mentioned.

METABOLIC PATHWAYS

The chain of chemical reactions involved in cell respiration is a typical example of a metabolic pathway. After entering the cell cytoplasm as a 6-carbon unit, glucose is split into two 3-carbon units (pyruvic acid). Each of these units then loses CO_2 to become a 2-carbon unit (acetic acid) which combines with coenzyme A to form *active acetate* (acetyl-coenzyme A), a step involving some of the B vitamins.

Although this two-carbon unit is a relatively small molecule it still has a high energy content. Total oxidation of a mole of acetic acid releases nearly 900 kilojoules.

$$\underset{\text{ACETIC ACID}}{CH_3COOH} + \underset{\text{OXYGEN}}{2O_2} \longrightarrow \underset{\text{CARBON DIOXIDE}}{2CO_2} + 2H_2O + \underset{\text{ENERGY}}{880\ \text{kJ}}$$

The process by which this energy is released in the cells was worked out by Krebs in Sheffield, England, earning him a Nobel prize, and named after him the Krebs' citric acid cycle. It is rather complex but the main steps may be represented as follows (Fig. 28.4.). There are many more steps in the cycle than have been shown in the above diagram. Each of the steps involves specific enzymes and coenzymes. These are located in the mitochondria in an orderly fashion. To facilitate these reactions the surface area within the mitochondria is very greatly

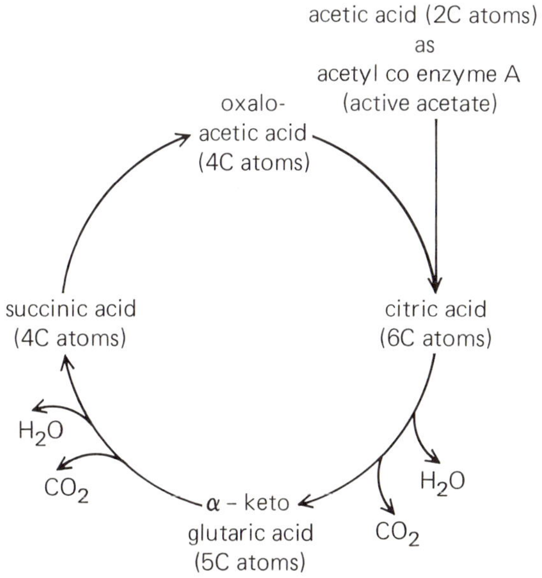

Figure 28.4
Kreb's citric acid cycle.

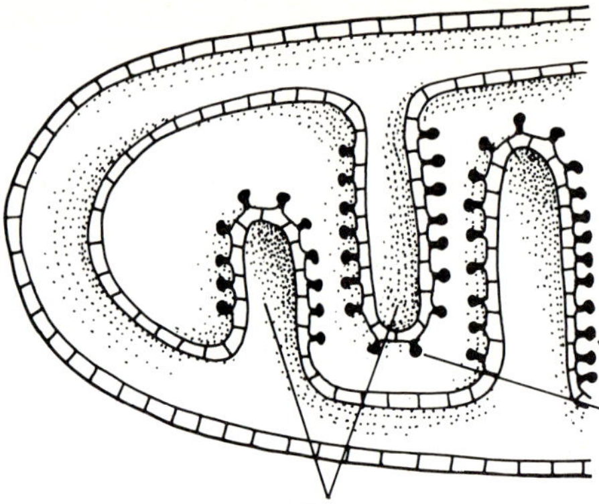

Figure 28.5
Part of mitochondrion. (See also Fig. 8.5(a) and (b).)

increased by partitions (cristae) incompletely dividing the lumen (Fig. 28.5).

The energy released at various steps of the citric acid cycle is used to convert ADP to ATP. In the cells of some tissues the mitochondria are lined up so that the ATP produced may readily be delivered to the organelles requiring energy, for example, the contractile fibers of muscle cells, the secretory portions of pancreatic cells and the midpiece of spermatozoa.

Many of the steps involved in metabolic pathways are reversible. Thus when energy is required, glycogen can be broken down into glucose. When there is more glucose than is needed for immediate needs, it can be converted with the aid of insulin into glycogen for storage until required, either in the muscles or in the liver.

Glucose, a 6-carbon unit, can also be converted into ribose, a 5-carbon unit. Ribose is an essential component of ribonucleic acid through which cell function is controlled and all hereditary characteristics are transmitted within the cell.

The pathways involved in the metabolism of fats and proteins each involve a whole series of specific enzymes. The first steps in the breakdown of both fats and proteins take place in the digestive tract. The fats are broken down into fatty acids and glycerol. In this form they are absorbed. Glycerol is a 3-carbon molecule (closely related to the trioses) and can be used by the cells for fuel without much chemical alteration. The fatty acids require considerably more modification before they can be used for fuel. The series of steps involved (β-oxidation) results in the formation of ketone bodies such as aceto-acetic acid (see Chapter 29, Fat). These can be converted into active acetate (acetyl coenzyme A) to produce energy through the citric acid cycle, provided that sufficient carbohydrate metabolism is also taking place. Otherwise ketosis will develop, as occurs in diabetes.

Proteins are broken down in the digestive tract into amino-acids. The amino-acids are absorbed to provide part of the body's amino-acid pool. A proportion of the amino-acids are used as fuel. By removal of the amino groups (de-amination) they can be converted to either carbohydrates or fatty acids and metabolized as such. Normally, however, a major part of the amino-acid pool is used for protein synthesis.

PROTEIN SYNTHESIS

The synthesis of proteins from the body's amino-acid pool is one of the most important energy-consuming processes in cell chemistry. Some of the amino-acids required can themselves be synthesized in the body. But there are a number of vital amino-acids which cannot be synthesized by the body and have therefore to be provided in the diet. They are known as the essential amino-acids. Most animal protein contains all the essential amino-acids. So meat is often referred to as first class protein. Much vegetable protein is lacking in one or more essential amino-acids and is therefore considered as second class protein. By careful choice of a vegetable diet it is possible to minimize the deficiencies; but man is by nature partly carnivorous and strict vegetarians can suffer deficiencies.

Protein synthesis requires ATP and specific nucleic acids to act as cofactors and templates. The templates which provide the pattern or jig for the construction of proteins consist of RNA (ribonucleic acid). The RNA, known as messenger RNA, derives its pattern from the DNA (deoxyribonucleic acid) of the chromosomes in the nucleus. The chromosomes carry the hereditary factors in the form of a chemical code in the DNA structure. The information is transcribed to messenger RNA. The RNA then leaves the nucleus, enters the cytoplasm and adheres to ribosomes where the protein synthesis is completed with incredible rapidity.

29

Body Fluids: Blood and Cerebrospinal Fluid

The body fluids on which chemical examinations are frequently undertaken are blood, cerebrospinal fluid, urine, and also fluids from various parts of the alimentary tract. Sweat may occasionally provide chemical information of diagnostic importance, for example, in fibrocystic disease of the pancreas. Blood and cerebrospinal fluid are discussed in this chapter; the other body fluids will be covered in Chapter 30. The pleural, peritoneal and joint cavities normally contain only sufficient fluid to provide lubrication between the surfaces but in certain pathological conditions the volume of fluid is increased and samples may then be submitted for laboratory examination.

BLOOD

Blood consists of plasma and cells. Plasma is water in which a number of chemical substances are dissolved or suspended. The cells present are red cells, erythrocytes; white cells, leucocytes; and platelets, thrombocytes. The chemical substances in plasma include salts, electrolytes; proteins, including antibodies; food substances; waste materials; and hormones. To obtain plasma, blood has to be collected into a container with an anticoagulant. Without an anticoagulant blood normally clots within 4 to 7 minutes and then gradually separates into blood clot and serum. Serum is chemically very similar to plasma but lacks certain clotting factors, notably fibrinogen.

Water

The volumes of the plasma, of extracellular water and of total body water may be significantly altered in dehydration due to burns and other causes of fluid loss (see Chapter 12), also in edema, obesity, and malnutrition. It is possible to measure these separate volumes by means of blood examinations, using dilution techniques. For example, a known volume of Evans blue (T1824) may be injected intravenously. A blood sample is then withdrawn, after allowing just sufficient time for mixing to occur. From a colorimetric examination of the plasma, the degree of dilution is estimated and the plasma volume calculated. In a normal adult it is about 3.5 liters. Blood-volume computers, which normally measure the dilution of a known amount of radioisotope, are programmed to give a readout of the volume of blood, extracellular fluid, and occasionally, of total body water.

Extracellular fluid volume may be estimated by using the radioisotope of a substance such as a chloride or bromide which readily diffuses into the extracellular tissue spaces. A sample of blood is then withdrawn as before; the degree of dilution is estimated from the radioactivity which is measured, for example, by a scintillation counter. To obtain the volume of total body water a dilution technique using heavy water, deuterium oxide D_2O, or radioactive water, tritium oxide T_2O, can be used but is expensive; urea is cheaper but less accurate.

The normal extracellular fluid volume is about 15 liters (20% of body weight) and the total body water about 50 liters (70% of body weight) for a man of 154 pounds (70 kg). Dehydration occurs when 5% of the body weight is lost as water, about 3 liters for an adult. Twice this loss may be fatal. Small infants are especially susceptible to dehydration; their kidneys are unable to concentrate urine to the same degree as adults and they have less reserve fluid.

Salts or Electrolytes

The blood electrolytes which are usually estimated are the cations, sodium $Na^{\oplus}$ and potassium $K^{\oplus}$, and the anions, chloride $Cl^{\ominus}$ and bicarbonate $HCO_3^{\ominus}$. Some aspects of extracellular and intracellular electrolytes have already been discussed (Chapter 13). The relative proportion of the various electrolytes in blood plasma can best be appreciated diagrammatically (Fig. 29.1); this method of representation was developed by

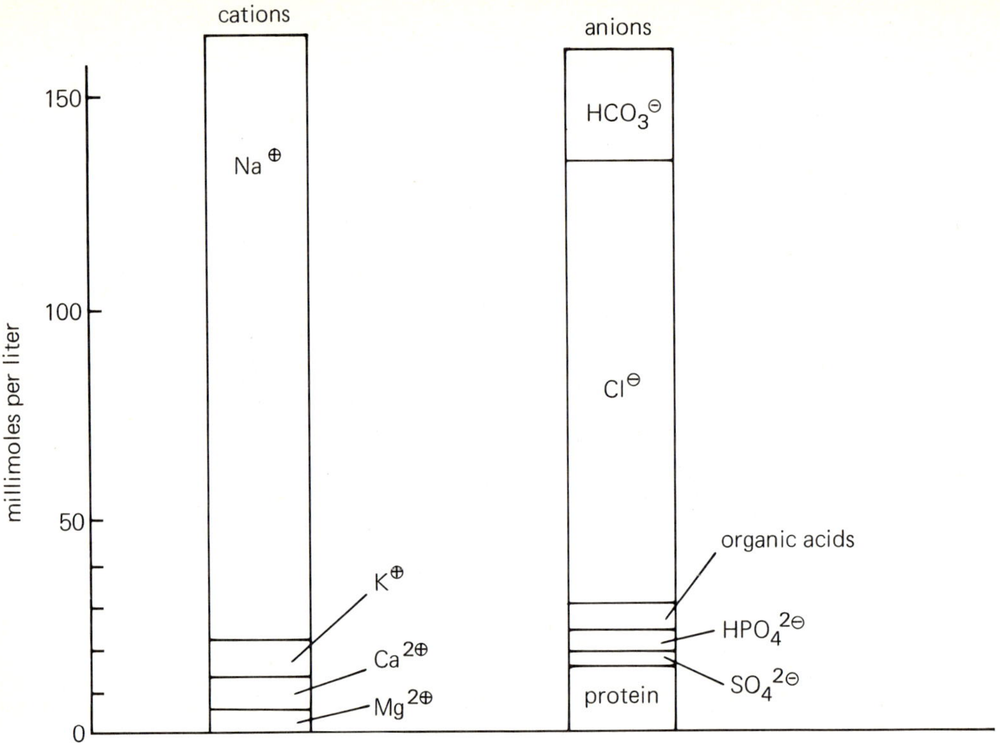

Figure 29.1
Representation of electrolytes in plasma (after Gamble).

Gamble who has greatly advanced the interpretation of electrolyte disturbance in various clinical conditions.

The electrolytes in extracellular tissue fluid are very similar to those in blood but virtually no proteins are present. The importance of electrolytes in maintaining the correct osmotic pressure in the blood has already been considered. The salts also play a vital role in maintaining blood pH by their buffering action.

Calcium $Ca^{2\oplus}$ and Magnesium $Mg^{2\oplus}$ ions in the blood affect muscle contractility and nervous irritability. Calcium is also essential for blood clotting. The normal serum calcium is 2.25 to 2.60 mmol/liter (9.0 to 10.4 mg/100 cm^3) and magnesium 0.6 to 1.0 mmol/liter (1.5 to 2.4 mg/100 cm^3). The serum calcium and magnesium levels fall in hypoparathyroidism and intestinal malabsorption.

Other metals in blood which may be estimated are iron, copper and zinc. The normal serum iron concentration of 1.1 to 3.2 mmol/liter (60 to 180 $\mu g/cm^3$) is usually reduced in iron deficiency anemia but a more constant finding in this condition is the raised iron-binding capacity above the normal range of 4.5 to 8 mmol/liter (250 to 450 $\mu g/cm^3$). The serum iron is raised in toxic and infective liver disease.

The normal serum copper is 12 to 25 μmol/liter (75 to 160 $\mu g/100\ cm^3$). It is raised in pregnancy, liver diseases and anemia of infection. It is reduced in nephrosis and kwashiorkor (protein starvation). Most of the copper is bound to a plasma protein called ceruloplasmin. In Wilson's disease there is a deficiency of ceruloplasmin: copper is deposited in the tissues, including the brain where it is one of the causes of Parkinsonism, and there is an increased excretion of copper in the urine.

Proteins

The proteins in the blood are usually estimated on the serum and so do not include fibrinogen which will be considered separately. The main serum proteins are albumin and globulin. Normally the total serum protein is 60 to 80 g/liter (6 to 8 $g/100\ cm^3$); albumin is 35 to 48 g/liter (3.4 to 4.8 $g/100\ cm^3$), the remainder being globulin. The importance of these proteins in maintaining osmotic pressure and preventing our tissues from becoming waterlogged has already been described (Chapter 14).

The serum proteins may be separated into a number of fractions by electrophoresis. The passage of an electric current causes the different components of the proteins to move at different speeds along a prepared filter paper (or other medium such as cellulose acetate strip) causing them to separate. The globulin is separated into three components called α, β and γ globulin (respectively, called alpha, beta, and gamma globulin).

The γ globulin includes many of the antibodies. During recovery from an infection there is a definite rise in the γ globulin level as a result of antibody formation. Deficiency of γ globulin, a condition known as agammaglobulinemia, results in a lowered resistance to infection due to the deficient antibodies.

The γ globulins are produced by plasma cells. Sometimes a tumor of the plasma cells occurs, called myeloma (also myelomatosis or multiple myeloma). The abnormal plasma cells of the tumor produce an abnormal γ globulin in considerable quantity. This can be clearly demonstrated by electrophoresis.

Since antibodies are globulins concerned with immunity they are now often called immunoglobulins. It is possible to produce antibodies to the immunoglobulins. By using this technique several different sorts of immunoglobulin have been recognized. That with the largest molecules, having a molecular weight of about a million, is called a macroglobulin or IgM, short for immunoglobulin M (M for macro, large); the normal serum level is 0.5 to 2.0 g/liter (50 to 200 $mg/100\ cm^3$). The γ globulin is called IgG, short for immunoglobulin G (G for gamma); the normal serum level is 6 to 16 g/liter (600 to 1600 mg/100 cm^3). Other immunoglobulins are IgA, IgD, and IgE.

By combining the use of antibodies with electrophoresis (immunoelectrophoresis) more than one hundred different proteins have been recognized in the blood. They include transferrin, the protein which transfers iron from gut to tissues, and ceruloplasmin, the protein which carries copper in plasma.

The plasma protein most immediately concerned with blood clotting is *fibrinogen*. The normal plasma fibrinogen is 2 to 4 g/liter (200 to 400 mg/100 cm^3). It is normally increased during pregnancy. It may occasionally drop to dangerously low levels following intrauterine fetal death. It has a rather long molecule but it is soluble. It is converted by thrombin into insoluble fibrin which precipitates in the form of numerous interweaving threads forming the blood clot:

$$\underset{\text{(SOLUBLE PROTEIN)}}{\text{fibrinogen}} + \text{thrombin} \longrightarrow \underset{\text{(INSOLUBLE BLOOD CLOT)}}{\text{fibrin}}$$

Normally blood does not contain thrombin. Otherwise we would clot spontaneously. Instead it contains an inactive precursor called prothombin. It requires thromboplastin (also called thrombokinase) in the presence of calcium ions to convert it to thrombin:

$$\text{prothombin} + \text{thromboplastin} + Ca^{2\oplus} \longrightarrow \text{thrombin}$$

Normally blood does not contain thromboplastin. Otherwise the prothombin present would be converted to thrombin and again we would clot spontaneously. So instead there are a number of factors which must interact before thromboplastin is formed. In order to start this interaction there must be damage either to tissue, which releases tissue thromboplastin, or to platelets, which are readily damaged if they touch a water-wettable surface.

If any one of the factors necessary for the formation of thromboplastin is missing from the blood there is an abnormal tendency to bleed (hemorrhagic diathesis). For example, in hemophilia there is a deficiency of a certain globulin. Since this globulin can cure the bleeding tendency in hemophilia it is called antihemophilic globulin. There is a similar condition called Christmas disease (Christmas was the name of the first patient in whom the disease was recognized) and the factor which is deficient is called Christmas factor. More than a dozen different factors have now been recognized.

Understanding the factors concerned in blood clotting has made the management of patients with hemophilia and similar haemorrhagic conditions much more successful. It also enables us to control undesirable clotting in blood vessels, for example, leg vein thrombosis with its often fatal hazard of pulmonary embolism, by the use of anticoagulants such as Warfarin (the same substance which is used as a rat poison).

Hemoglobin

This is a conjugated protein (see Chapter 25) that is synthesized in the red bone marrow and forms the main constituent of red blood cells. The red cell envelope prevents it from escaping into the bloodstream where its presence in large amounts would be toxic. It combines with oxygen to form oxyhemoglobin which is bright red and is readily reduced to provide oxygen for the tissues. The estimation of hemoglobin in blood provides a measure of its oxygen-carrying capacity. This is reduced in anemia. The internationally accepted figure for 100% hemoglobin in blood is 14.6 g/100 cm^3. It is higher in men (14–17 g/100 cm^3) and lower in women (12–15 g/100 cm^3). At birth it is about 20 g/100 cm^3 but rapidly falls during the first few weeks of life to about 15 g/100 cm^3.

Food substances

Sugar. The main blood sugar is glucose in the form of dextrose (D-configuration glucose). The normal fasting level for whole blood is 3.6 to 5.6 mmol/liter (65 to 100 mg/100 cm^3). For plasma it is slightly higher, being 4.2 to 6.0 mmol/liter (75 to 107 mg/100 cm^3). The blood sugar level is normally prevented from rising too high by the hormone insulin, secreted by the islet cells of the pancreas. A high fasting blood sugar level is characteristic of diabetes mellitus and it may require insulin injections to keep the blood sugar down to normal levels. In mild diabetes an appropriate diet may provide sufficient control, thus eliminating the need for injections. Recently new agents have become available which lower the blood sugar and can be taken by mouth.

An overdose of insulin will cause an excessively low blood sugar with resulting weakness and mental confusion. Lower blood sugar levels, namely, less than 2.2 mmol/liter (40 mg/100 cm^3) cause unconsciousness. This is called hypoglycemic coma or insulin coma. It is characterized by sweating and deep breathing. If treated promptly by an intravenous injection of dextrose solution, recovery of consciousness is very rapid. Untreated, it can be fatal.

Sometimes the islet cells of the pancreas are overactive (as in insulinoma) and produce too much insulin. The effect is just the same as an overdose of insulin and is known as spontaneous hypoglycemia (Gk. *hypo*, under, + Gr. *glukos*, sweet or sugar, + Gr. *aemia*, blood state). Apart from surgical removal of the overactive islet cells it can be treated by glucagon, a substance which can be given by mouth and which counteracts the effect of insulin by raising the blood sugar. Glucagon, like

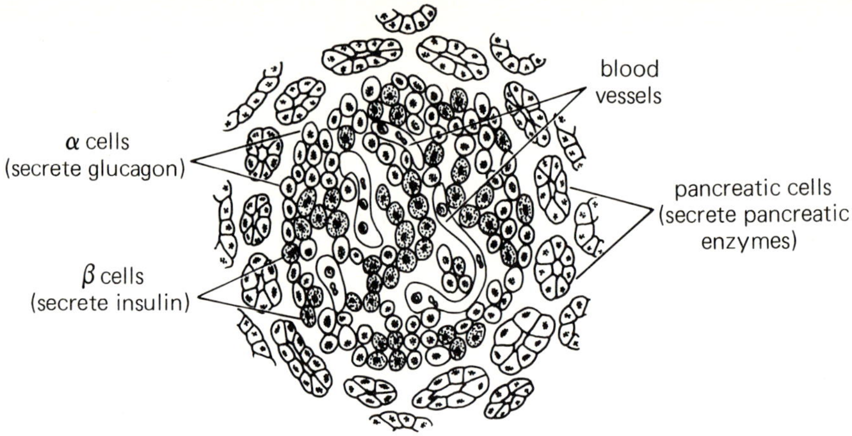

Figure 29.2
Islet of Langerhans: many small islands of such tissue are present in the pancreas; each is well supplied with blood vessels into which the hormones insulin and glucagon are secreted.

insulin, is a hormone secreted by islet cells of the pancreas. The cells producing glucagon are called α cells. Those producing insulin are called β cells (Fig. 29.2).

The weakness which occurs if a meal is too long delayed is also due to low blood sugar. It is counteracted by eating, the most rapidly absorbed and metabolized form of food being carbohydrate, especially in the form of glucose.

The *glucose tolerance* test provides a means of studying the body's ability to stabilize blood sugar. The blood sugar levels are measured after fasting and at half-hourly intervals after ingesting 50 g of glucose (Fig. 29.3).

Normally, a peak blood sugar level of less than 10 mmol/liter (180 g/100 cm^3) is reached within about half an hour and returns to a normal fasting level in $1\frac{1}{2}$ hours. In diabetes mellitus, not only is the $\frac{1}{2}$-hour blood sugar higher than normal but also the level continues to rise, reaching an abnormally high peak or plateau at about $1\frac{1}{2}$ hours. This high blood sugar or hyperglycemia (Gr. *hyper*, over) is characteristic of diabetes. In suspected spontaneous hypoglycemia the glucose tolerance test is usually continued for 5 hours in order that the fall in blood sugar to very low levels, for example, below 3.6 mmol/liter (65 mg/100 cm^3) may not be missed.

Fat

Most of the fat in blood is present as neutral fat or as fatty acids. The fatty acid level is highest in the portal vein where the fatty acids which

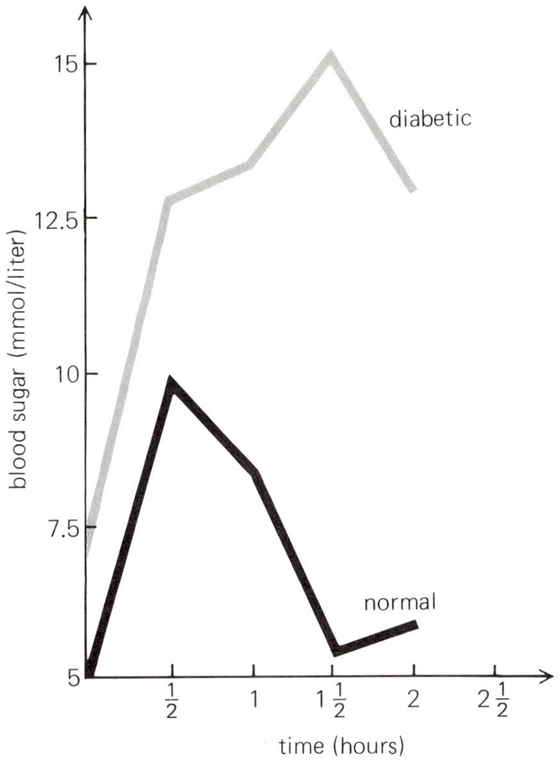

Figure 29.3
Glucose tolerance test. (Modified from *Special Tests and Their Meanings*, 9th edition, by D. M. D. Evans. Courtesy Faber & Faber, London.)

have been absorbed following digestion are transported to the liver to be metabolized. They can be seen under the microscope as small fatty droplets, *chylomicrons.*

Fatty acid metabolism and glucose metabolism are linked together at several points, so that if less glucose is available, the energy requirements can be made up by metabolizing more fatty acids. However, when fatty acids are being metabolized faster than glucose, intermediates in the fatty acid metabolism build up causing the formation of excess ketone bodies.

The ketone bodies are

acetone	CH_3COCH_3
aceto-acetic acid	CH_3COCH_2COOH
β-hydroxybutyric acid	$CH_3CH(OH)CH_2COOH$

The tissues can use these as a source of energy but only to a limited extent. Traces of ketone bodies can be found in normal nonfasting blood.

Fasting for 2 to 3 days causes this level to be greatly increased, fifty times or more. In diabetes there is a breakdown in glucose metabolism. So the ketone bodies accumulate in the blood causing ketosis, excessive ketones, and acidosis, excessive acid; the ketone bodies are passed in the urine.

Excessive fat in the blood is known as *lipemia*. It has been shown that there is a relationship between lipemia and coronary artery disease. So the management of heart disease due to coronary atheroma involves limiting certain types of fat in the diet, particularly those with a high content of cholesterol and lecithin.

Another form of blood fat (lipid) is *cholesterol*. The normal serum cholesterol is 3.6 to 5.7 mmol/liter (140 to 220 mg/100 cm^3). The level is increased during pregnancy, also in diabetes mellitus, myxedema and cretinism, chronic nephritis with edema, and biliary tract obstruction. It is reduced in thyrotoxicosis and chronic wasting diseases. The steroid structure of cholesterol has been previously described (Chapter 20, Fused rings). It is the raw material for the liver's synthesis of bile salts (Chapter 27, Bile) which aid the digestion and absorption of fat. It is also the substance from which the adrenal cortex (the outer part of the suprarenal gland) synthesizes the steroid hormones, such as cortisone, and from which the gonads synthesize the sex hormones.

The blood also contains fat compounds with a very complex structure called phospholipids, for example, lecithin. These are essential components of normal cell structure and are particularly abundant in cells of the central nervous system (see the last section in Chapter 23).

Amino-Acids

All protein in the diet is digested to the amino-acid stage before it is absorbed. The amino-acids absorbed into the blood stream form part of the general amino-acid pool. This pool enables the structure and functional efficiency of the essential organs to be maintained. It provides the raw material from which all enzymes are synthesized, also the plasma proteins, including the various antibodies, and those hormones which are either proteins or derived from protein. Insulin is an example of a hormone which is a protein.

Vitamins

Several vitamins are present in the blood in measurable quantities, for example, vitamin B_{12}, folic acid, and vitamin C. Of these it is the estimation of vitamin B_{12} which is probably of most value.

The amount of *vitamin B_{12} (cyanocobalamin)* normally present in blood serum is 110 to 440 pmol/liter (150 to 600 pg/cm^3). This is such a low concentration that ordinary chemical methods are not sufficiently sensitive to measure it. So instead a technique called microbiological assay is used. This involves using an organism such as euglena gracilis which requires vitamin B_{12} in order to grow and multiply. By providing it with all other food requirements except vitamin B_{12}, its rate of growth then becomes dependent on the amount of vitamin B_{12} present. It thus provides a very sensitive way of measuring the reduced vitamin B_{12} in serum which is characteristic of pernicious anemia (see Chapter 19, Vitamin B_{12}).

The amount of *folic acid* in the blood is measured in a similar way, using a different organism, for example, streptococcus fecalis. The normal folic acid level is more than 5.7 μmol/liter (2.5 μg/cm^3). It is reduced in megaloblastic anemia of pregnancy, particularly if there is more than one fetus to drain the mother's reserves, and in intestinal malabsorption.

Although *vitamin C (ascorbic acid)* is reduced below the normal level of 20 to 100 μmol/liter (0.4 to 2.0 mg/100 cm^3) in scurvy a much more reliable method of detecting this condition is the ascorbic-acid saturation test (see Chapter 30, Vitamin C).

Waste materials

Urea. Urea

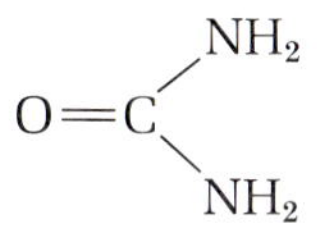

is the main end product of protein breakdown. It is derived chiefly from food protein but also partly from the breakdown of tissue protein. It is the substance which is most frequently estimated to check kidney function. The normal blood urea is 2.5 to 7.5 mmol/liter (15 to 45 mg/100 cm^3). Healthy kidneys have considerable reserve powers of excretion. Even removal of the whole of one kidney may not cause the blood urea to rise. If kidney function is impaired to a sufficient degree to cause the blood urea to rise above the normal range the condition is known as uremia. In elderly people, however, a slight rise in the blood urea may not be very significant.

Uremia may be due to factors operating before the blood reaches the kidney (prerenal), factors in the kidney itself, renal, and factors preventing the urine leaving the kidney, post-renal.

1. *Pre-renal:* The blood pressure may not be sufficient to provide an adequate filtration force through the renal glomeruli. Fall in

blood pressure may be sudden, as in shock, for example, from hemorrhage or gradual, as in some forms of heart failure.

2. *Renal:* The function of the kidney cells may be impaired, either permanently, as in certain forms of nephritis, or temporarily, as in acute pyelonephritis due to an infection which can be treated or the reversible shock nephrosis resulting from the prerenal factors described in (1).
3. *Postrenal:* The outflow from the kidneys may be obstructed in the kidney pelves, ureters, bladder or urethra. Obstruction may be from outside, for example, compression by pregnancy; in the wall of the organ, for example, a tumor; or within the lumen, for example, a stone (calculus).

In each case the excretion of urea and other waste products is prevented, leading to progressive uremia. Unless treated, this is likely to be fatal when the blood urea reaches 50 to 100 mmol/liter (300 to 600 mg/100 cm^3). Treatment involves removing the cause, for example, obstruction or infection and/or assisting kidney function by renal dialysis or kidney transplant.

Uric acid. Uric acid is the main end product of nucleoprotein breakdown. Like urea it is derived partly from tissue breakdown and partly from the diet. Its formula is more complicated than that of urea and it has a double ring structure:

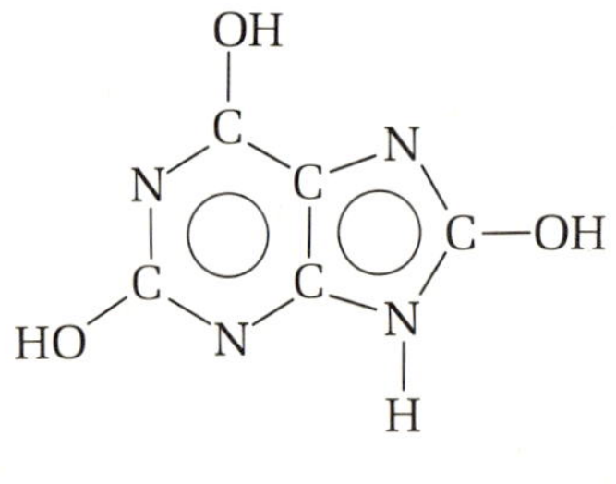

There are a group of substances having a similar structure, known collectively as purines. Foods rich in nucleic acid such as liver, kidney and pancreas (sweetbread) are rich in nucleoproteins, the main source of purines. Meat extracts, tea, coffee and cocoa also provide purines.

In gout there is an abnormal purine metabolism associated with arthritis. Too much uric acid is produced and at the same time the excretion of uric acid by the kidneys is impaired; so the blood level rises and urates (salts of uric acid) are deposited in the joint tissues.

Normally the blood uric acid is 0.1 to 0.35 mmol/liter (1.5 to 6 mg/100 cm^3) for women and 0.15 to 0.4 mmol/liter (2.5 to 7 mg/100 cm^3) for men. In gout the level may rise to 0.6 mmol/liter (10 mg/100 cm^3). The blood level is also increased in other conditions where the excretion of uric acid is impaired or its formation is increased. Uric acid excretion is impaired when there is uremia in acute or chronic nephritis, also in lead poisoning. Its formation is increased in leukemia.

CEREBROSPINAL FLUID (CSF)

Cerebrospinal fluid is extracellular fluid which provides a liquid cushion round the central nervous systems. Normally it is a clear liquid. It is formed within the cavities (ventricles) of the brain and passes out through small apertures (foramina) in the roof of the fourth ventricle to enter the subarachnoid space which surrounds the brain and spinal cord (Fig. 29.4). Samples of cerebrospinal fluid are usually obtained by in-

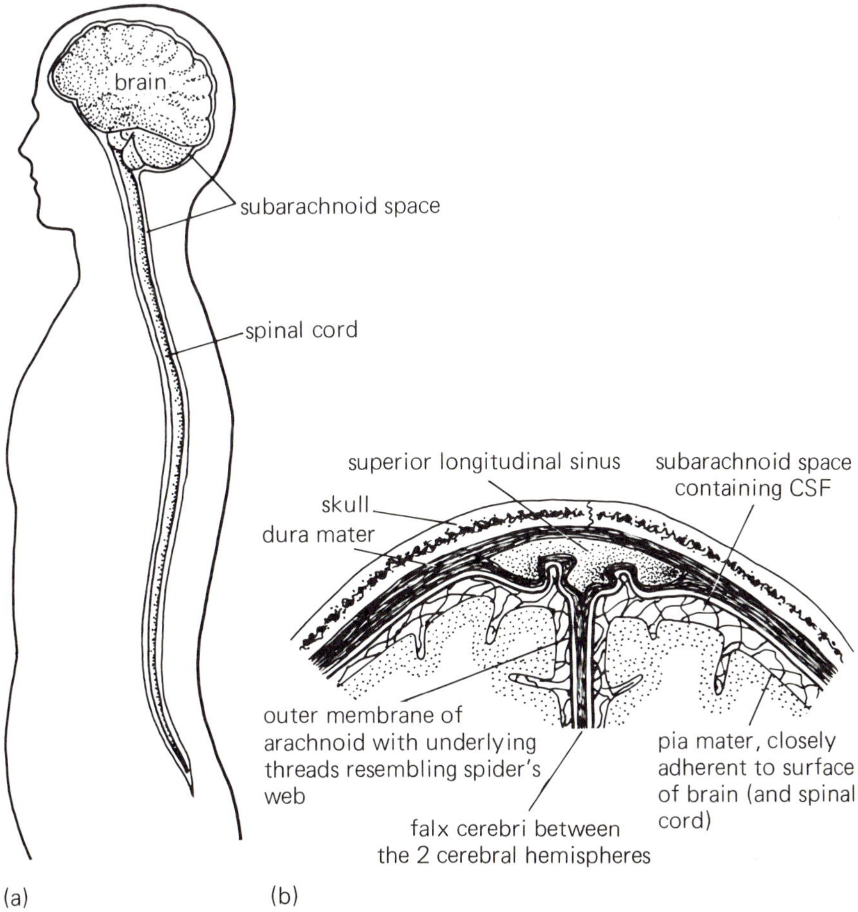

Figure 29.4
(a) Subarachnoid space containing cerebrospinal fluid cushion surrounding brain and spinal cord; (b) subarachnoid space showing relationship to brain surface, membranes and skull.

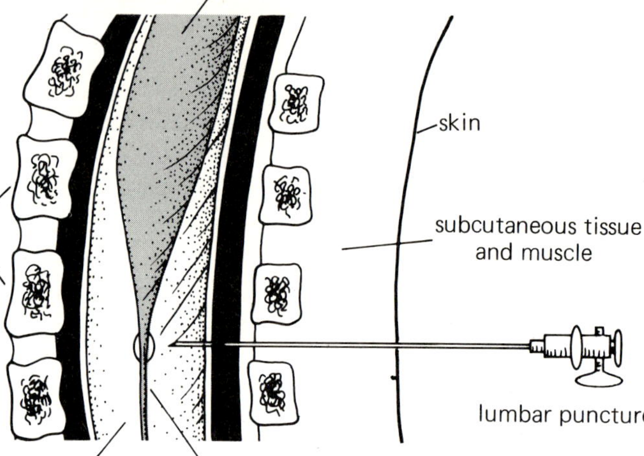

Figure 29.5
Lumbar puncture.

serting a needle with full aseptic precautions, through the skin over the lumbar spine into the subarachnoid space of the spinal canal just below the lower end of the spinal cord (Fig. 29.5).

The tissue which forms the cerebrospinal fluid is called the choroid plexus. Its main component is a network of capillaries. The process by which the cerebrospinal fluid is formed from the blood is evidently more than that of simple filtration through the walls of the capillaries. Some substances such as urea readily pass from blood to CSF and vice versa. So do some ions, for example, $Na^{\oplus}$, $Cl^{\ominus}$, but others such as $SO_4^{2\ominus}$, $NO_3^{\ominus}$ and $I^{\ominus}$ can pass through into the CSF only with difficulty, although entering the extracellular tissue spaces in the remainder of the body quite readily.

Glucose

Glucose is one substance whose concentration of 2.8 to 5.0 mmol/liter (50 to 90 mg/100 cm^3) in the CSF is normally very similar to that in blood. In diabetes there is a rise in the CSF glucose level corresponding to that in the blood but lagging slightly behind it; in hypoglycemia there is a corresponding fall in the CSF glucose level.

In meningitis due to meningococcal and other septic infections the CSF glucose falls to a very low level. In tuberculous meningitis it falls below 2.8 mmol/liter but does not usually reach such low levels as in septic meningitis.

Chloride

Normal CSF contains more chloride than blood plasma. In adults it is 123 to 133 mmol/liter (96 to 106 mmol/liter in blood plasma). In children the normal CSF range is 110 to 130 mmol/liter. A fall or rise in plasma chloride produces a corresponding change in the CSF level.

In acute meningitis and to an even greater degree in tuberculous meningitis there is a fall in the CSF chloride level. This may be at least partly due to vomiting with resultant chloride loss.

Protein

The normal concentration of protein in the CSF is 0.15 to 0.40 g/liter (15 to 40 mg/100 cm^3), much less than that in the blood plasma (60 to 80 g/liter or 6 to 8 g/100 cm^3). In all inflammations (septic, tuberculous, syphilitic, fungal, and viral) involving the meninges and brain the CSF protein is raised, probably due to leakage through capillary walls damaged by the infection. The CSF protein is also raised in many other organic diseases of the central nervous system, including tumors.

Urea

The concentration of urea in the CSF is the same as that in blood, namely, normally 2.5 to 7.5 mmol/liter (15 to 45 mg/100 cm^3). If there is kidney failure of sufficient degree to cause uremia the rise in CSF urea is practically the same as that of blood urea.

30

Body Fluids: Urine, Alimentary Tract Fluids, Sweat, Tears, and Milk

URINE

Urine is an aqueous solution of inorganic and organic substances. It is formed by the kidneys in two stages. The first stage is filtration of the fluid portion of the blood through the glomeruli (Figure 14.7, Chapter 14). The second stage is selective reabsorption by the tubule cells of those chemicals required by the body, including water (Figure 14.8, Chapter 14) Analysis of urine provides information of much value in the diagnosis and management of illness. It is analyzed as a ward or clinic procedure and also, usually in greater depth, as a laboratory procedure.

Ward or clinic examination of urine

This includes general appearance, volume and specific gravity together with simple chemical tests for reaction, albumin, sugar, ketones and blood.

General appearance

1. *Color:* A dilute urine is usually pale and strawcolored. A concentrated urine is dark. Acid urines tend to be darker than alkaline ones. Blood may cause urine to be red, brown or smoky.

   In obstructive jaundice the presence of bile pigment in the urine causes it to be dark orange (bilirubin), changing to green (biliverdin) due to oxidation by air on standing. Such urine readily forms foam of a similar color on shaking, due to lowering of the surface tension by bile salts. There are a number of different substances which may give urine a characteristic color, for example, red from eating beetroot.
2. *Turbidity:* Freshly passed normal urine is clear but may become cloudy due to the presence of phosphates which precipitate on cooling to room temperature. Phosphate cloudiness disappears on addition of acid but cloudiness due to pus and/or bacteria does not. All urines become cloudy on standing due to bacterial growth and alkaline salt formations.

Volume. Measurement of the volume of each urine specimen passed is necessary for conditions in which the fluid balance is affected, for example, operations, shock, dehydration, edema, kidney failure, steroid therapy, and so forth.

Specific gravity. Specific gravity is measured by placing the urine in a cylinder and floating a urinometer in it (Fig. 30.1). The specific gravity is indicated by the level of the urine surface against the calibrations on the neck of the urinometer. Normally the first morning urine has a specific gravity of 1.015 to 1.020. Healthy kidneys respond to water deprivation by producing concentrated urine of small volume and high specific gravity, about 1.030. They respond to a large fluid intake by producing a dilute urine of large volume and low specific gravity, about 1.002. Failing kidneys, unable to make these adjustments, produce urine with a specific gravity in the region of 1.010 regardless of fluid intake.

Routine chemical tests. Until recently, examining the urine for reaction, albumin, sugar, ketones and blood involved a series of different tests. It is now possible to undertake all these tests simultaneously by means of a single disposable test strip, for example, Labstix (Ames Co.). The test area of the strip is dipped into fresh, clean urine (free from antiseptics, detergents or acid). It is immediately withdrawn and the edge tapped against the side of the container to remove excess urine. The colors of the test areas are then compared with the corresponding color charts at the times specified, namely, immediately, 10 seconds, 15 sec-

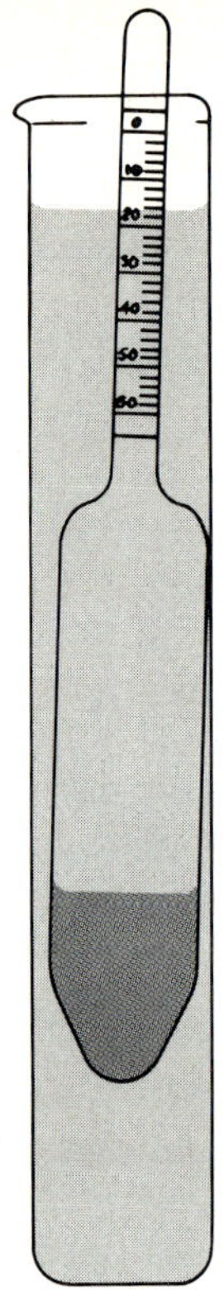

Figure 30.1
Urinometer.

onds and 30 seconds. Any abnormal result is checked by laboratory examination.

NOTE. The only sugar which is estimated by this method is glucose. Other sugars such as lactose and galactose are not detected so that in young children, where it is important not to miss an inborn error of metabolism, a more general method of sugar testing should also be used, for example, clinitest (Ames Co.).

1. *Reaction:* Normal urine is usually slightly acid with a pH of about 6.0 (normal range pH 4.8 to 7.5). Abnormally acid urine occurs in acidosis, for example, from diabetic ketosis, starvation or severe diarrhea. Alkaline urine occurs in alkalosis, for example, from severe vomiting or excess alkali administration for peptic ulcer and in many urinary infections due to the organisms which split urea and release ammonia:

$$O{=}C(NH_2)_2 + H_2O \longrightarrow 2NH_3 + CO_2$$

UREA AMMONIA CARBON DIOXIDE

Some carbonic acid is formed by the carbon dioxide in solution but it is not very soluble and so the main effect on the pH is produced by the formation of ammonium hydroxide:

$$NH_3 + H_2O \longrightarrow NH_4OH$$

AMMONIA AMMONIUM HYDROXIDE

2. *Albumin:* Albumin in the urine (albuminuria) generally implies leaking of plasma albumin through the renal glomerulus. It does not necessarily mean kidney disease although this is an important cause. It can occur temporarily in otherwise healthy young people, later disappearing completely. Albuminuria also occurs following violent exercise, during fevers, in certain forms of chemical poisoning and as a result of urine contamination by pus, blood or seminal fluid.
3. *Glucose:* Glucose is one of the substances which pass through the renal glomerulus into the Bowman's capsule, is reabsorbed by the renal tubules and so returned back to the blood. The amount of glucose in the urine is normally so small that it is not detectable by the tests used clinically. Readily demonstrable amounts of glucose are only found in the urine when the blood level exceeds a certain threshold value, normally 10 mmol/liter (180 mg/100 cm^3). In untreated diabetes mellitus the blood level will exceed this value after a carbohydrate meal, with the result that glucose appears in the urine (glycosuria). The higher the blood glucose level the greater the glycosuria. It should be noted that even when sufficient insulin has been given to cause the blood sugar to return to normal there may still be residual urine in the bladder containing glucose.

   Glycosuria does not necessarily mean that diabetes is present. It can occur, for example, if there is a low renal threshold so that glucose is passed in the urine when the blood glucose is still within normal limits. This condition is known as *renal glycosuria.*
4. *Ketones:* The ketone bodies are a normal product of fatty acid metabolism and can be used by the tissues as a source of energy, provided there is sufficient glucose combustion as well. When there is deficient glucose metabolism, as in starvation and dia-

betes mellitus, the ketone bodies accummulate in the blood and are excreted into the urine. So ketonuria with glucosuria is characteristic of diabetic coma and ketonuria without glucosuria occurs in starvation.

Laboratory examination of urine

The routine urine analysis includes all the procedures described above together with microscopy and bacteriology. In addition to cells and casts, microscopy often reveals crystals of chemical substances in the urine for example, uric acid, urates, phosphates, oxalates (especially after eating rhubarb, Fig. 30.2), cystine or sulfonamides. Special examinations for various chemical substances in the urine include the following.

Amino-acids. Normal urine contains only small amounts of certain amino acids, for example, glycine and glutamine. Abnormal amounts and types of amino acid are found in liver failure and in a number of congenital metabolic diseases. They are separated and identified by chromatography. One or two drops of the amino acid mixture are spotted onto a sheet of chromatography paper about 1 cm from the bottom. The sheet is then suspended in a tank so that the paper below the spot is im-

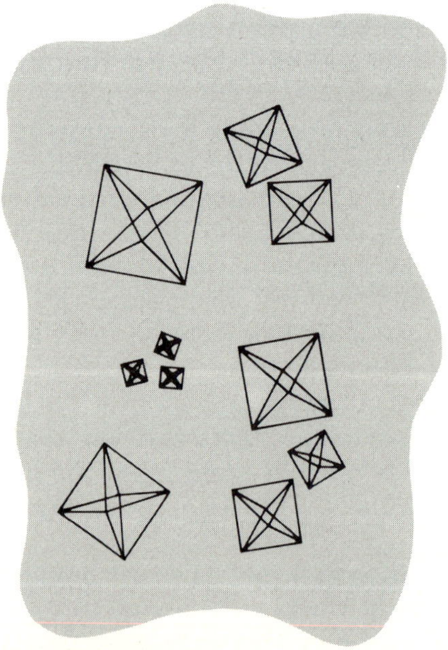

Figure 30.2
Oxalate crystals in urine.

mersed in a suitable solvent. The solvent rises, by capillary action, carrying or eluting the various amino acids at different speeds up the paper. In this way, they are separated into individual spots. These are colorless but are rendered blue by spraying with a solution of ninhydrin. The distance traveled by an amino acid divided by the distance traveled by the solvent is known as the R_f value of the acid and is characteristic of it. Amino acids can therefore be identified by their R_f values.

Bile pigments and salts. The bile pigment bilirubin is altered in the intestine to a series of compounds generally referred to together as urobilinogen (stercobilin). Normally some of this is reabsorbed and then reexcreted by the kidney. In the urine the urobilinogen gradually changes to urobilin, these pigments being a major factor in giving urine its characteristic color. In hemolytic jaundice their excretion is increased, up to 10 mg daily, and in obstructive jaundice it is reduced, usually to less than 0.3 mg daily.

Bilirubin itself only appears in the urine when there is an appreciable amount of conjugated (direct) bilirubin present in the plasma. So it is found in obstructive jaundice and in the early stages of infective hepatitis but is not found in hemolytic jaundice.

Obstructive jaundice is caused by something blocking the bile duct, for example, gallstone or tumor, so that bile pigments and salts cannot escape into the intestine. As a result these substances enter the blood and are filtered out by the kidney glomeruli into the urine. The pigments make the urine very dark and the salts make it froth readily. Their presence can be confirmed chemically.

Calcium. On an average diet, 2.5 to 7.5 mmol (100 to 300 mg) of calcium are excreted in the urine daily. It is considerably more if the diet is rich in milk and cheese. Urine calcium is low in rickets, due to deficiency of vitamin D needed to absorb calcium by the gut. It is similarly low in intestinal malabsorption which can cause adult rickets. It is raised in hyperparathyroidism and myelomatosis.

Creatine. Creatine

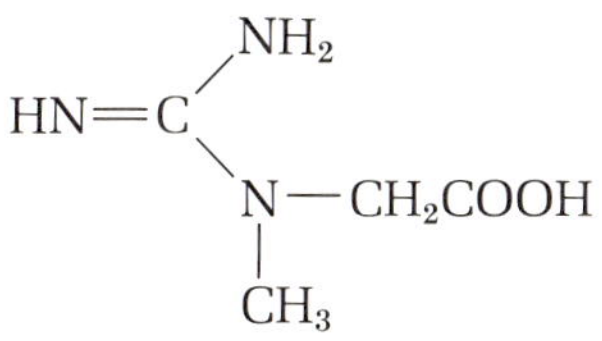

is present in muscle in the form of an energy-rich phosphate. It provides the energy reserve for replenishing the ATP, adenosine triphosphate, required for muscular contraction. Normally, it does not leave the muscle except in the form of its breakdown product creatinine. Consequently, normal urine contains little or no creatine. Some may be found during menstruation, pregnancy and childhood, also following athletic activity. Urine creatine is considerably increased in conditions involving muscular wasting.

Creatinine. After creatine has been used by the muscles it is converted into creatinine,

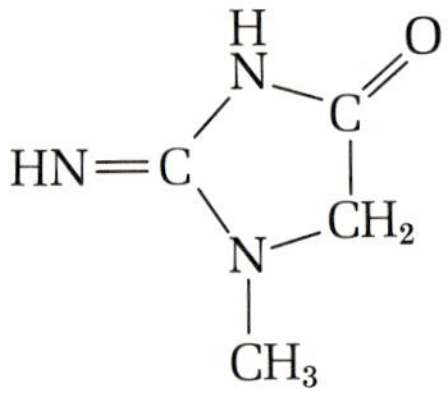

its anhydride (creatine from which one molecule of water has been removed). Normally 10 to 30 mmol (1.1 to 3.4 g) of creatinine is excreted daily. For any individual the amount of creatinine excreted each day is so constant that it can be used to check that 24 hour urine collection is complete when a series of such samples is required for chemical or microbiological assay.

Drugs. Urine levels, for example, for alcohol, barbiturates, and so forth, are often higher than those in blood. Although they provide less accurate information than the corresponding blood levels in cases of suspected intoxication, they may yet provide valuable information, particularly if no blood sample is available for any reason.

Electrolytes. Estimation of urine chloride, sodium and potassium can be of considerable value in conditions where the electrolyte balance is disturbed (see Chapter 13).

Hormones. Increased or diminished production of a hormone in the body is often reflected by a corresponding alteration in the amount of the hormone or its breakdown product excreted in the urine. This applies particularly to hormones produced by the adrenal glands, gonads and the pituitary gland.

The *adrenal medulla* produces the hormones *adrenaline* and *noradrenaline*. Adrenaline and related substances are called catechol amines. In a tumor of the adrenal medulla (pheochromocytoma) there is an excess of catechol amines, both in the blood, causing high blood pressure, and also in the urine. Their estimation in a 24 hour urine sample enables this condition to be diagnosed. Normally the excretion of catechol amines is less than 150 μg/day. Rather similar substances, for example, 5-hydroxytryptamine or serotonin, are produced in excess by carcinoid tumors. These occur in the appendix but may also arise in other sites.

The *adrenal cortex, testes,* and *ovaries* produce steroids of different types. The 17 ketosteroids in the urine come from both adrenal cortex and testes. The 17 ketogenic steroids come from the adrenal cortex only. Both types of steroid are reduced in Addison's disease which is due to failure of the adrenal cortex. Tumor or overgrowth of the adrenal cortex causes virilism (masculinization) in females and boys, with increased steroid formation and excretion. The ovarian steroid hormones are estrogen and progesterone. Their breakdown products in the urine can now be estimated.

The *pituitary gland* produces *gonadotrophins* that stimulate the gonads. During pregnancy gonadotrophin is also produced by the chorion (the outer membrane around the fetus). This chorionic gonadotrophin plays a major part in maintaining pregnancy, preventing menstruation from taking place. Pregnancy tests are based on demonstrating the increased chorionic gonadotrophin present in the urine.

The *pituitary gland* also produces *follicle stimulating hormone* (FSH) that stimulates the ovary to form a Graafian follicle and initiates the menstrual cycle. In pituitary deficiency no FSH can be detected in the urine.

Porphyrins. Porphyrins (Gr. *porphyros,* purple) are not breakdown products of hemoglobin as was thought previously, but rather intermediates in hemoglobin synthesis. The structure of the porphyrin skeleton is illustrated in Fig. 30.3.

Normally the amount excreted in the urine is very small, 15 to 25 μg. daily. If red cells are being destroyed excessively in the body, as in hemolytic anemia, the body tries to compensate by greatly increasing the rate of hemoglobin synthesis. Excess porphyrin is produced and so more is excreted. In a group of diseases with a hereditary factor called the porphyrias, there is excessive excretion of the related substances uroporphyrin and porphobilinogen. In an acute attack there may be sufficient in the urine to turn it purple.

Protein. Albumin is the protein most commonly found in the urine. Its significance has already been discussed. Bence Jones proteins appear in the urine in myelomatosis. They are subunits of IGG and having a relatively small molecular size can pass through the glomerular filter. They have the peculiar property of precipitating when the urine is heated to 50°C and redissolving when the temperature reaches 80°C. By using antibodies to Bence Jones proteins, they can be detected in much smaller amounts; that is, much earlier in the disease than they can by heating.

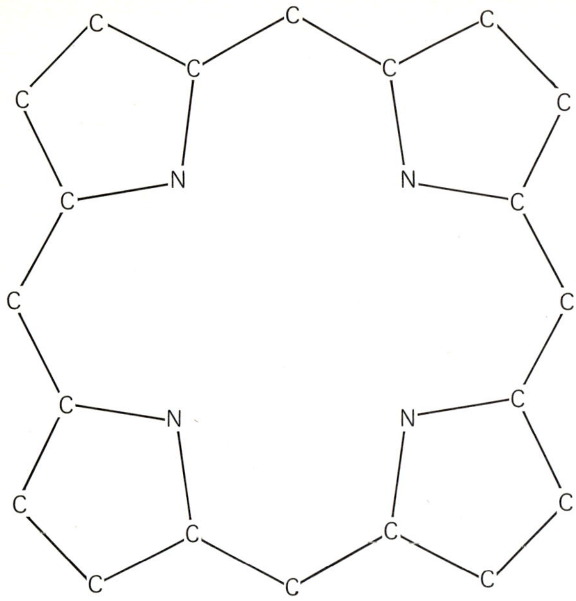

Figure 30.3
Porphyrin skeleton.

Urea. The urea

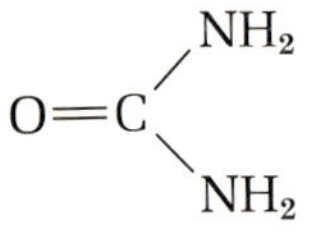

in urine is usually estimated as part of a renal function test. A high concentration of urea indicates that the kidney is functioning effectively. An average concentration of urea over the day is about 2% with a total daily excretion of 0.2 to 0.5 mol (12 to 30 g).

Vitamin C, or Ascorbic acid. Ascorbic acid

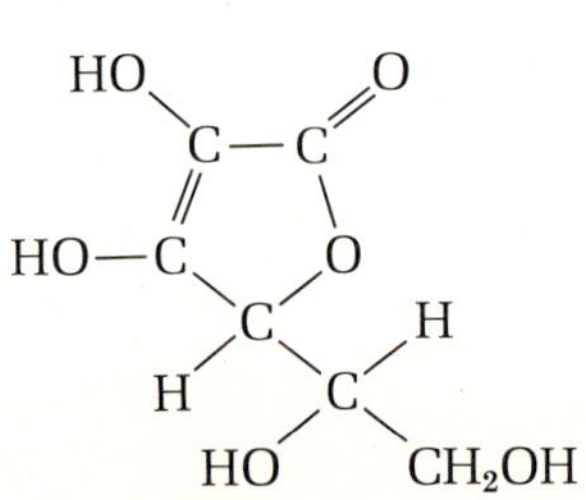

is a strong reducing agent. This is because there is a double bond between the two carbon atoms (top left of formula) that is readily converted to a single bond with the loss of 2 hydrogen atoms.

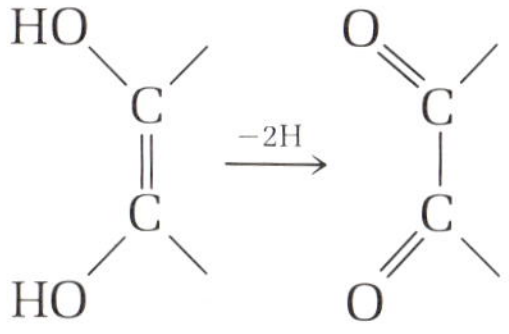

It is believed that this property enables vitamin C (or its derivatives) to act as a redox catalyst inside cells.

Deficiency of ascorbic acid causes scurvy, characterized by bleeding gums, skin thickening with enlarged hair follicles, poor healing of wounds and, in infants, painful hemorrhages under the periosteum usually near the end of long bones. Suspected ascorbic acid deficiency can be detected by measuring the amount of ascorbic acid excreted in the urine between the fourth and sixth hour after taking a large measured dose of this vitamin. This is the *ascorbic acid saturation test.* In scurvy practically none of the vitamin C is excreted as it is rapidly taken up by the depleted tissues.

ALIMENTARY TRACT FLUIDS

The fluids of the alimentary tract include saliva, gastric juice, bile, pancreatic juice and intestinal juice or succus entericus. Their chemistry has already been considered in Chapter 27. Chemical examination of the feces can also provide valuable information.

Feces

Occult blood. Large amounts of blood in the feces (melaena) can usually be recognized visually. Small amounts of blood may not be evident and are therefore referred to as occult (hidden). Its presence can be demonstrated by simple ward tests, for example, Hemoccult (Smith Kline Diagnostics). The patient must be on a meat-and-fish–free diet. Vigorous tooth-brushing and menstruation can give misleading positive results for occult blood.

The finding of occult blood may provide the clue to an iron deficiency anemia which does not appear to respond to iron. It is most commonly due to a bleeding peptic ulcer, (gastric or duodenal), but may also be due to tumors or other diseases affecting any part of the alimentary tract.

Fat. The amount of split and unsplit fat in the feces can be estimated in the laboratory. In pancreatic disease there is often, but not always, an increase in the amount of unsplit fat (the lipase in the intestinal juice can usually split only about 70% of the fat). In intestinal malabsorption it is the absorption rather than the splitting of the fat which is affected so that the feces contain an increased amount of split and unsplit fat.

Muscle Fibers. In pancreatic disease the stools may show a considerable increase in the number of undigested muscle fibers. However, this is not a constant finding because muscle fibers are digested partly by the acid gastric juice with its pepsin and partly by the pancreatic juice with its trypsin (activated by intestinal juice), the breakdown process being completed by the peptidases in the intestinal juice.

Trypsin. In children the examination of feces for trypsin is a useful guide to the possible presence of pancreatic disease. If fresh feces are sent to the laboratory the presence and amount of trypsin can be demonstrated by its ability to digest protein, for example, gelatin on an exposed X-ray film. Trypsin is diminished or absent in pancreatic disease.

SWEAT

Even when the skin appears dry, insensible perspiration is taking place. This sweat is practically pure water and its evaporation serves to cool the body. Its volume can amount to 600 to 700 cm^3/day. The volume of sweat and its salt content may be considerably increased by temperature, exertion and emotion. In cystic disease of the pancreas there is an abnormally high salt content of sweat and tears, the sodium content usually exceeding 60 mmol/liter. This finding is of diagnostic value. It also makes its victims unduly susceptible to the effects of hot weather, resulting in collapse due to sodium loss. This may be corrected by the administration of sodium chloride. Similarly, miner's or stoker's cramps, may be prevented by including salt in the drinking water.

TEARS

The fluid which bathes the cornea is produced partly from the tear (lachrymal) glands and partly from accessory sebaceous (Meibomian and Zeis) glands. When tears are rapidly produced, as in crying, the electrolyte content and pH are very similar to that in blood plasma. When tear production is slow, as in the normal state, evaporation results in the tear fluid becoming hypertonic. The concentration of protein in tears (6 to 8 g/liter) is about a tenth of that in plasma. Tears contain an enzyme lysozyme also found in nasal mucus, sputum, gastric secretion, milk, eggwhite, and in tissues, which hydrolyzes components of bacterial cell walls and so helps to protect the cornea from infection.

Table IX. CHEMICAL COMPOSITION OF HUMAN AND COW'S MILK

MILK	TOTAL SOLIDS (g/liter)	PROTEIN (g/liter)	FAT (g/liter)	CARBOHYDRATE (g/liter)	ASH (g/liter)
human	125	15	35	70	2
cow's	130	35	47	45	7.5

MILK

Milk is an almost complete natural food. It contains proteins, fats, carbohydrates, minerals and some vitamins. It is deficient in iron, also in copper and in vitamins C and D.

The general composition of milk is similar for all mammals but there are significant differences between the various species. Those that grow fastest have the most protein in their milk. A rabbit doubles its weight in 6 days and its milk has a protein content of 104 g/liter. A cow doubles its weight in 50 days and has a milk protein content of 35 g/liter. A human takes 180 days to double its weight and its milk protein content is 15 g/liter.

Humans so often drink cows' milk that a comparison of their chemistry is of more than academic interest (Table IX).

The main differences are the higher protein and ash content in cow's milk and the higher carbohydrate concentration in human milk. To make cow's milk suitable for human infants it needs diluting with about an equal volume of water and then having lactose added to bring it up to approximately the human level.

The ash contains the electrolytes, that is, calcium, magnesium, sodium, potassium, phosphate, and chlorine ions. The proportions of each are very similar in both human and cow's milk, but the total amounts are more than three times greater in cow's milk. Cow's milk is about the ideal nutritional source for calcium and phosphate.

The protein in cow's milk is 80% casein. It forms the curds when milk is acidified, the remaining liquid being known as the whey. In human milk only 40% of the protein is casein, the remining 60% being whey proteins. Human and bovine caseins are very similar. Another important milk protein is β lactoglobulin, present in both human and bovine whey; it is ideal for supporting growth in young animals.

The fats in milk are mainly triglycerides. They are dispersed in a fine emulsion making them digestible even by gastric lipase. Carbohydrate in milk is essentially lactose (milk sugar). It is a disaccharide. On digestion it is hydrolyzed into galactose and glucose.

31

Radioactivity

Radioactivity was one of those chance discoveries made by prepared minds. X Rays had already been discovered in 1895 by the German physicist von Röntgen. One of the properties of these rays is that they cause certain uranium salts to fluoresce (give out light). The French physicist Henri Bequerel in 1896 attempted to reverse this process; that is to produce X rays by exposing a uranium salt to bright sunlight. After exposure, he placed the salt under a well-wrapped and therefore light-sealed photographic plate in a darkroom. After several hours the plate was darkened which made him suppose at first that exposure of the uranium salt to light had caused it to produce X rays which had penetrated the wrapping. However, Bequerel later noticed to his surprise that the same effect was produced by a sample of the salt which had been kept in a dark cupboard for a number of days. So he realized that the penetrating radiation was not produced by the action of light on the uranium salt but was a property of the salt itself. He called this spontaneous emission of radiation radioactivity. It was soon shown that it was the uranium and not the other elements in the salt which was responsible for the emitted radiation.

More radioactive elements were quickly discovered. Marie Curie found that the element thorium was radioactive and she also noticed that pitchblende, a uranium ore, was far more radioactive than its uranium content indicated. She reasoned that it must contain at least one unknown element which was far more radioactive than uranium. In 1898, she and her husband, Pierre, isolated polonium (named after her native Poland) from pitchblende and later that year they obtained from the same ore a second, far more radioactive element, radium. The Curies and Bequerel were jointly awarded the Nobel prize for physics in recognition of their discoveries. A few years later, Marie Curie received an unprecedented second Nobel prize, this time for chemistry.

At the turn of the century an intense effort was put into the study of radioactivity by some of the world's leading physicists and chemists. Three types of radiation were discovered and were called alpha α, beta β and gamma γ rays, this being the order of their penetrating power. Alpha rays do not penetrate beyond the superficial layers of the skin. Nonetheless, it is dangerous to ingest even traces of long-acting α emitters because they accummulate in the bones and other tissues. Beta rays penetrate further, but still fail to get through the skin and reach the internal organs. Gamma rays are extremely penetrating and cannot be absorbed by a lead shield several centimeters thick. They penetrate the body, affecting internal organs and bones.

Alpha rays were shown to be streams of positive particles and these α particles were identified as helium nuclei $He^{2\oplus}$ (Table X). Beta rays were shown to be streams of negative particles and these β particles turned out to be electrons. Gamma rays do not consist of electrically

Table X. SUBATOMIC PARTICLES

	PARTICLE	ATOMIC MASS UNITS	CHARGE	COMMENT
	electron	1/1840	−1	known as a beta (β) particle when emitted by a radioactive substance.
	neutron	1	0	electrically neutral particle
	proton	1	+1	nucleus of hydrogen atom
	deuteron	2	+1	nucleus of deuterium atom
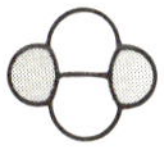	alpha (α) particle	4	+2	nucleus of helium atom

charged particles. They are identical with short wavelength X rays and belong to the broad spectrum of radiation which includes ultraviolet, visible and infrared light and radiowaves.

All these types of radiation emitted by radioactive elements have sufficient energy to knock out electrons from molecules which they hit, thus producing positive ions. It is this ionizing ability that enables us to detect these radiations. In a cloud chamber, water vapor condenses on the ions producing tell tale vapor tracks of the rays (Fig. 31.1). The reverse process occurs in a bubble chamber where a liquid, usually liquid hydrogen or liquid propane, forms bubbles of vapor around the ions (Fig. 31.2). In a Geiger counter (Fig. 31.3) positive ions, formed by α and β particles, are attracted to a negative electrode giving an electrical signal. This produces a click in a loudspeaker for each α and β particle entering the counter. It is not particularly effective in detecting γ rays.

The ionizing power of α, β and γ rays makes them both a health hazard and a useful tool in the treatment of cancer. Molecules in the living cell are ionized by the rays and the ions have different chemical properties from the parent molecules. The chemistry of the cell is therefore altered, resulting either in cell damage or cell destruction. Cells undergoing division are particularly susceptible to damage affecting the chromosomes, with the result that they lose the capacity to divide. This is fine if the cells are malignant but distinctly hazardous if they are healthy dividing cells. So organs in which there is normally active division, particularly the bone marrow and gonads, are at the greatest risk. These radiations can in fact induce cancer in healthy cells. The photographs in Fig. 31.4 show some of the physical effects of radiation on survivors of the atomic bombing of Hiroshima in 1945. X Rays are also ionizing radiation and many of the pioneer radiographers suffered a life of agony, terminated by cancer, as a result of over exposure to X rays (Fig. 31.5).

Figure 31.1
Tracks made in a cloud chamber. (Photographed by P.M.S. Blackett, the Cavendish Laboratory. Reproduced with permission.)

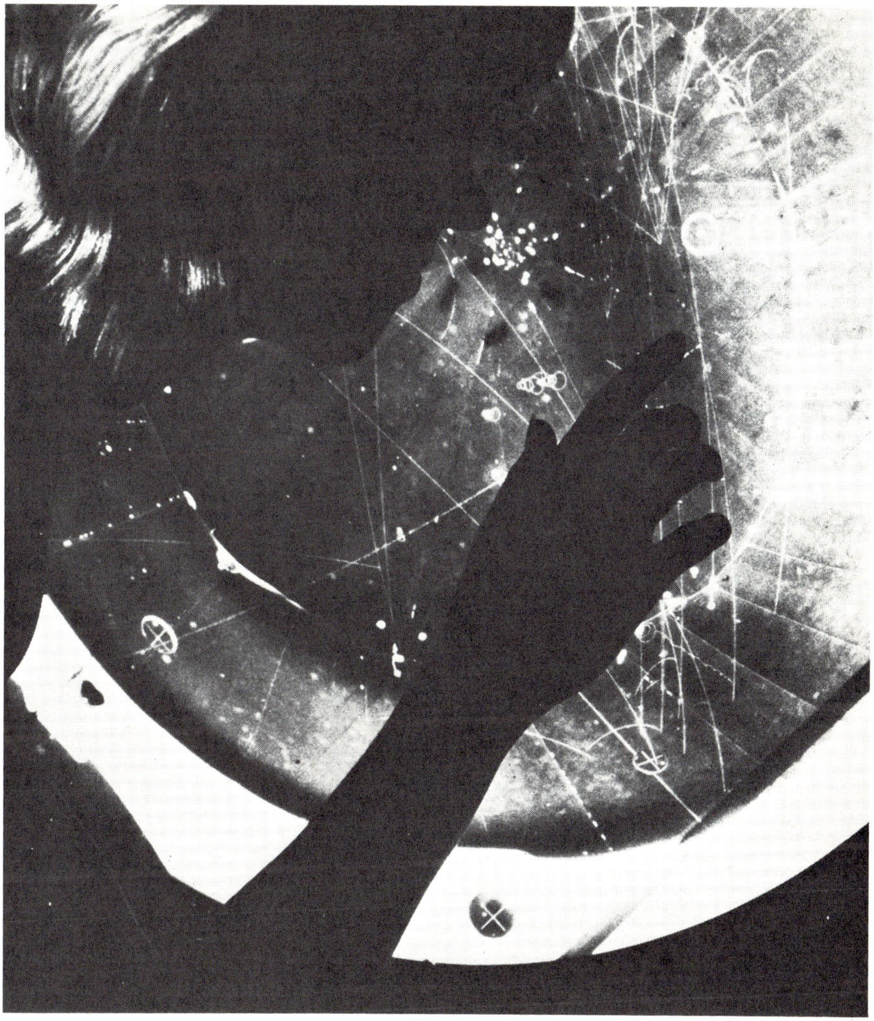

Figure 31.2
Bubble chamber tracks being examined by a technician. (Courtesy ERDA, Argonnes National Lab.)

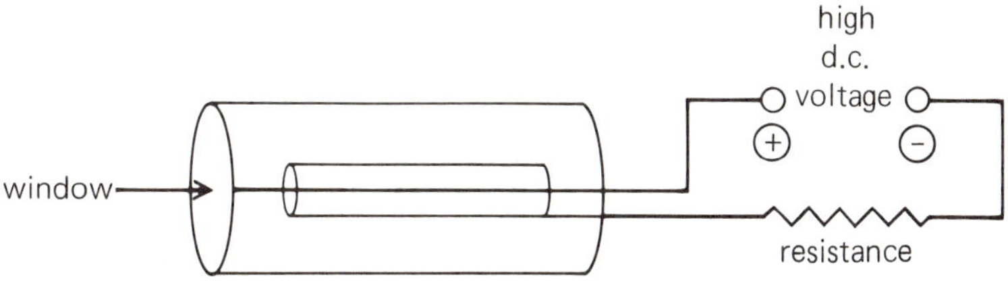

Figure 31.3
Geiger counter. Radiation passes through the window and ionizes the gas, usually argon, in the tube.

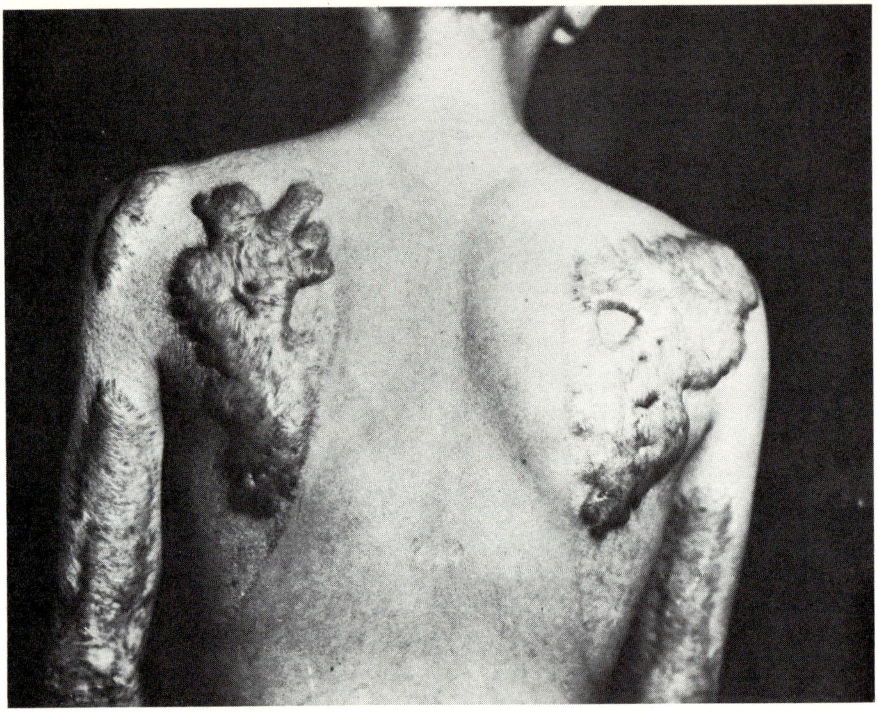

Figure 31.4
Atomic bomb survivor with scars of burns sustained at the time of the atomic bomb explosion at Hiroshima. (Courtesy Wide World.)

The harmful effects of radiation can be mitigated by injecting cysteine or glutathione as soon as possible after exposure. These are reducing agents; they give up electrons to the positive ions and so reduce them to their original molecular state.

The dose of radiation is measured in units called *rads*. One rad is the quantity of radiation which will result in an energy absorption of 0.01 joules per kilogram (or using abbreviations 0.01 J/kg) of the irradiated material. Numerically, the rad is very similar to the Röntgen which measures X-ray dosage. Irradiation of tissue with 1 Röntgen of X radiation leads to energy uptake of 0.0097 J/kg. A single dose of 800 rads is almost certainly lethal, producing nausea and vomiting after 2 hours, followed by a few days respite, after which vomiting returns accompanied by diarrhea and inflammation of the throat. This gives way to fever and rapid emaciation which is caused by the damaged stomach lining being unable to absorb food. Death occurs within two to four weeks of irradiation. A considerable proportion of people receiving a single dose of 600 rads would die within a few months. Doses below 400 rads should not prove fatal provided there are no complications. How-

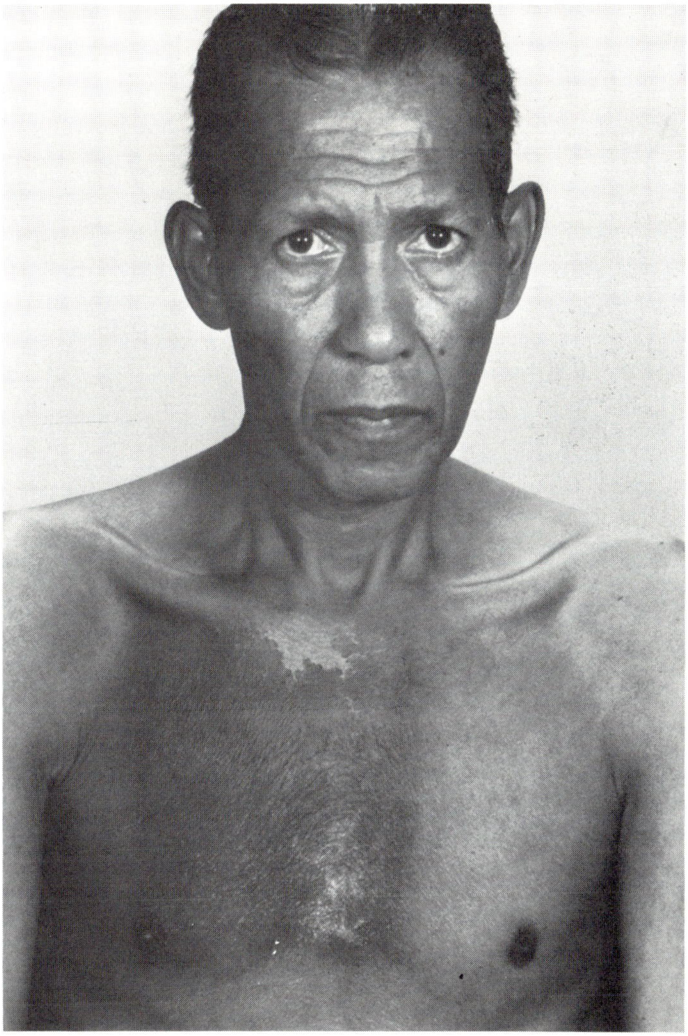

Figure 31.5
One type of radiation burn resulting from overexposure to X rays. (Courtesy Rotker, Taurus Photo.)

ever, repeated small doses have a cumulative effect and the accepted level to which hospital staff and other workers can be exposed, without harm, is 0.1 rads per week.

EMISSION OF α, β, AND γ RAYS

How do we account for the emission of α, β, and γ rays? Rutherford and Soddy made the startling discovery that a radioactive element was transmuted, that, is, changed into another element, called the daughter element when it emitted radiation. This was the alchemists' dream come true; a dream which Dalton's atomic theory had banished, a century earlier, to the realm of pure fantasy. The daughter element may itself be radioactive and change into another element and this, too, may be radioactive, so that a series of radioactive elements may be produced from the original parent. The series ends when a nonradioactive (stable) element is formed. In nature, there are three such series. One starts with uranium-238, ^{238}U another with its isotope uranium-235, ^{235}U and the third with thorium-232, ^{232}Th. Each of these series ends with a stable isotope of lead. The isotope ^{206}Pb terminates the ^{238}U series, ^{202}Pb the ^{235}U series and ^{208}Pb the ^{232}Th series.

Rutherford and Soddy deduced that when one element changes into another, its nucleus emitted either an α or a β particle. Radium-226 is an α emitter and its daughter element is the radioactive rare gas radon-222. The disintegration of the ^{226}Ra nucleus may be represented by the following equation:

$$\underset{\text{RADIUM-226}}{{}^{226}_{88}\mathrm{Ra}} = \underset{\text{RADON-222}}{{}^{222}_{86}\mathrm{Rn}} + \underset{\substack{\alpha\text{-PARTICLE} \\ \text{(HELIUM NUCLEUS)}}}{{}^{4}_{2}\mathrm{He}}$$

The upper figures are the mass numbers and the lower ones the atomic numbers. Clearly, if an α particle containing 2 protons and having a mass of 4 units, is ejected from a radium nucleus containing 88 protons and having a mass of 226 units, then the daughter nucleus must contain $88 - 2 = 86$ protons and have a mass of $226 - 4 = 222$ units. You can see therefore, that in this type of equation both the mass numbers and the atomic numbers must be balanced. You may recall from Chapter 4 that an element is characterized by the number of protons in its nucleus. A nucleus containing 86 protons is a radon nucleus.

The disintegration of thorium-234 is an example of β emission.

$$\underset{\text{THORIUM-234}}{{}^{234}_{90}\mathrm{Th}} = \underset{\text{PROTOACTINIUM-234}}{{}^{234}_{91}\mathrm{Pa}} + \underset{\substack{\beta\text{ PARTICLE} \\ \text{(ELECTRON)}}}{{}^{0}_{-1}\mathrm{e}}$$

The subscript number -1 in ${}^{0}_{-1}e$ indicates that the electron has a charge of -1 and the zero that it has virtually no mass. A β particle is emitted from the nucleus because a neutron breaks up into an electron, β particle, and a proton, hydrogen ion $H^{\oplus}$, which remains in the daughter nucleus

$${}^{1}_{0}n = {}^{1}_{1}H + {}^{0}_{-1}e$$

The daughter nucleus therefore contains one more proton than its parent. In this particular radioactive change the daughter nucleus has 91 protons and this makes it a protoactinium nucleus. The α or β emission is usually accompanied by the emission of γ radiation.

A radioactive change is quite different from an ordinary chemical change. The most obvious differences are the emission of radiation and the formation of a new element. Other important differences are that a radioactive change proceeds at a rate which is independent of temperature and pressure and of how the radioactive element is chemically combined. Also, the energy liberated in a radioactive change is far greater than in a normal chemical reaction.

HALF-LIFE

It has been shown experimentally that the time taken for half the nuclei of a radioactive element to disintegrate is a constant and independent of the temperature, the pressure and the initial quantity of the element. Thus 10 g of radium is reduced to 5 g in the same time that 4 g is reduced to 2 g. This time interval is called the half-life of the element. Each radioactive element has its own characteristic half-life and so can be identified by it. Half-lives range from millions of years to a fraction of a second. The isotope ^{238}U for example, has a half-life of 4,500 million years, while that of ^{212}Po, polonium-212, is less than a millionth of a second.

The fact that half the nuclei of a radioactive element decay in a fixed time implies that radioactivity is governed by the laws of chance, since each nucleus has a 50/50 chance of disintegrating during the half-life period and these odds remain fixed throughout successive half-lives.

32

Applications of Radioactivity

RADIOACTIVE DATING

Age of the Earth

The age of the earth can be estimated by measuring the fraction of ^{238}U atoms which has decayed in rocks containing this isotope. This is done by determining the ratio of ^{206}Pb atoms to ^{238}U atoms in the rocks. It has been found to be approximately 1:2. Now, ^{238}U decays extremely slowly to produce a ^{206}Pb atom. So it can be inferred that one out of every three ^{238}U atoms, that is one-third, has decayed since the earth was born. Scientists know how long it takes for any given fraction to disintegrate and so the age of the earliest rock can be calculated. It turns out to be about 3500 million years.

Age of organic material

The technique known as radiocarbon, ^{14}C, dating is used for estimating the age of this kind of material. The method is based on a determination of the ratio of ^{14}C atoms to ^{12}C atoms in the material.

The radioisotope, ^{14}C is formed in the atmosphere by the bombardment of nitrogen nuclei with neutrons from cosmic rays, that is, radiation from outer space.

$$\underset{}{^{14}_{7}N} + {^{1}_{0}n} = \underset{\text{RADIOCARBON}}{^{14}_{6}C} + \underset{\text{PROTON}}{^{1}_{1}H}$$

The radiocarbon is formed as fast as it decays so that the ratio of ^{14}C to ^{12}C in atmospheric carbon dioxide is constant. Thus plants, which breathe in carbon dioxide during photosynthesis, will have this same constant $^{14}C/^{12}C$ ratio in their tissues, as will plant-eating animals and humans. Now, when the living organism dies its ^{14}C atoms are not replenished and so the $^{14}C/^{12}C$ ratio decreases. A knowledge of this ratio enables its age to be calculated because it is known how this ratio varies with time. The age of the Dead Sea scrolls was established in this way. The method is limited to material which is less than 50,000 years old because after this time virtually all its ^{14}C will have decayed.

BOMBARDMENT REACTIONS

The first artificial transmutation of an element was achieved by Rutherford in 1919 when he bombarded nitrogen nuclei with α particles. He did this by simply passing α particles from ^{214}Bi through nitrogen gas. The bombardment resulted in the formation of the ^{17}O isotope of oxygen and also of hydrogen:

$$^{14}_{7}N + {^{4}_{2}He} = {^{17}_{8}O} + {^{1}_{1}H}$$

This reaction set the stage for atomic or nuclear fission (atom-splitting) which was first accomplished in 1932 by Cockcroft and Walton when they bombarded lithium nuclei with fast-moving protons. These had sufficient energy to split the lithium nuclei into helium nuclei.

$$^{7}_{3}Li + {^{1}_{1}H} = 2\,{^{4}_{2}He}$$

The splitting of ^{235}U furnishes the enormous energy of the atomic bomb (see below).

In the following year 1933, Irene and Frederic Joliot, the daughter and son-in-law of Marie and Pierre Curie, bombarded aluminum with α particles and changed it into an isotope of phosphorus, namely ^{30}P, which was radioactive and therefore called radiophosphorus

$$^{27}_{13}Al + {^{4}_{2}He} = {^{30}_{15}P} + {^{1}_{0}n}$$

This was the first time that radioactivity had been produced artificially. Radioisotopes of all the elements have now been made. A number of them have medical uses.

Many different particles have been used as projectiles for bombarding nuclei. Protons, deuterons (or heavy-hydrogen or deuterium nuclei $^{2}_{1}H^{\oplus}$) and α particles all have the disadvantage of being positively charged and consequently repelled by positive nuclei, especially those of heavy elements which have a large positive charge. They are therefore only effective when moving at high speed. Machines such as cyclotrons and linear accelerators are used to accelerate these particles prior to nuclear bombardment.

The problem of repulsion does not arise with neutrons, as they are electrically neutral. They are therefore effective in bringing about nuclear reactions at relatively low speeds. Every known nucleus, apart from that of hydrogen, is changed when bombarded with neutrons.

In an atomic bomb, the ^{235}U nuclei are split by neutron bombardment. Every smashed nucleus emits a few neutrons, each of which, provided it does not escape from the surface of the metal, will split another ^{235}U nucleus producing more neutrons. Thus, a chain reaction is set up in which the number of neutrons multiplies very rapidly so that in less than a millionth of a second an enormous number of nuclei are split. Now, when a ^{235}U nucleus breaks up in this way a small proportion of its mass is destroyed. The total mass lost by all the ^{235}U nuclei in an atomic bomb explosion is about 600 g. If you bear in mind that the destruction of only 1 g of matter produces as much energy as the explosion of 33 tons of TNT (trinitrotoluene), we can appreciate the tremendous forces released in an atomic explosion.

Einstein derived the following simple relationship between the energy E produced when a mass m is destroyed

$$E = mc^2$$

where c is the velocity of light equal to 186,000 mi/sec, a fantastic speed. (Note that mc^2 means in mathematics $m \times c \times c$). You can see, therefore, from Einstein's equation that an enormous amount of energy is produced by the destruction of only a small amount of mass.

No special source of neutrons is needed in the atomic bomb to start the reaction, as there are always enough present in the air. One might well ask, in that case, why it does not explode spontaneously. The reason is that if the ^{235}U is below a certain critical size, too many neutrons escape from its surface for the chain reaction to develop. In the atomic bomb two pieces of uranium, each less than the critical size, are kept apart until the bomb is fired. An explosive charge then drives the pieces together to form a single lump of uranium, which exceeds the critical size.

In the hydrogen bomb, *fusion* (or joining) of nuclei occurs. Two deuterium nuclei fuse to form a helium nucleus:

$$^{2}_{1}H + ^{2}_{1}H = ^{4}_{2}He$$

The small mass loss in this reaction provides the energy for the explosion. This same reaction is the source of solar energy, as we saw in Chapter 11.

In nuclear power stations and nuclear submarines, energy is obtained in a nuclear reactor using the same reaction that occurs in the atomic bomb. How is the reaction controlled in these reactors? The secret is to lower cadmium rods into the ^{235}U as and when required. The cadmium absorbs the neutrons produced in the chain reaction which is consequently slowed down. The energy of the nuclear reaction appears as heat which is used to boil water and the steam is used to drive the electrical generators, and also, in the case of submarines, the propellers.

Nuclear reactors are also employed to produce radioisotopes for medical and other purposes. Various elements, for example, sodium, phosphorus, and cobalt are subjected to neutron bombardment inside the reactor and this changes them into their radioisotopes.

$$^{23}_{11}Na + ^{1}_{0}n = ^{24}_{11}Na$$

RADIOSODIUM

$$^{31}_{15}P + ^{1}_{0}n = ^{32}_{15}P$$

RADIOPHOSPHORUS

$$^{59}_{27}Co + ^{1}_{0}n = ^{60}_{27}Co$$

RADIOCOBALT

New elements, which do not occur in nature, were made in this way. Element 93, neptunium, and element 94, plutonium, were made by bombarding ^{238}U, with neutrons in a nuclear reactor. Unlike its sister isotope ^{235}U, it is nonfissionable and instead of exploding, it combines with the neutron to produce ^{239}U,

$$^{238}_{92}U + ^{1}_{0}n = ^{239}_{92}U$$

which rapidly undergoes β emission forming neptunium-239 ^{239}Np,

$$^{239}_{92}U = ^{239}_{93}Np + ^{0}_{-1}e$$

This isotope of neptunium has a half life of only 2.3 days and so it changes quickly into plutonium-239 ^{239}Pu, β particles (electrons) again being emitted

$$^{239}_{93}Np = ^{239}_{94}Pu + ^{0}_{-1}e$$

The isotope ^{239}Pu has a half life of about 24,000 years and like ^{235}U is a fissionable isotope. Both are used as fuels in nuclear reactors. Breeder reactors produce more plutonium from ^{238}U than they use as fuel; hence, the name breeder. This is just as well because supplies of ^{235}U fuel are very limited.

SOME MEDICAL USES OF RADIOISOTOPES

Research and investigation of disease

The radioactive and the stable isotopes of a particular element have virtually the same chemical properties. It is this fact which makes radioisotopes so useful. A minute and consequently safe amount, of a radioisotope with a reasonably short half life can be mixed with its stable sister isotope, enabling the path of the element through the human or animal body to be traced with a detector such as a Geiger counter. This radioactive tracer technique is frequently employed in medicine.

Radiosodium, ^{24}Na, is used to locate obstructions in the blood circulatory system as in the leg arteries causing intermittent claudication (limping). The patient swallows some ordinary sodium chloride $^{23}NaCl$ which contains a little radioactive $^{24}NaCl$. This is absorbed into the bloodstream and its circulation is traced with a Geiger counter so that any obstruction is easily located. Overactive and underactive thyroid glands can be detected by administering radioiodine, ^{131}I, orally in the form of sodium iodide. This is taken up by the thyroid which converts the iodine into thyroxine which passes into the blood. If the gland is overactive the iodine is taken up by the thyroid very rapidly and radioactive thyroxine appears in the bloodstream too quickly. If it is underactive the iodine is only taken up by the thyroid very slowly: the appearance of radioactive thyroxine in the blood is slower and less than normal, much of the radioiodine being lost in the urine. Other applications of radioisotopes in medicine are described under tritium (Chapter 11) and water (Chapter 29).

Radiotherapy

Radiocobalt, ^{60}Co, is now widely used instead of the more expensive radium. For the more accessible tumors the radiocobalt may be implanted directly into the malignant tissue in the form of needles, or into an immediately adjacent body cavity in the form of tubes. For less accessible tumors an apparatus such as that shown in Fig. 32.1 may be used. A high intensity source of radioactivity is provided by radiocobalt con-

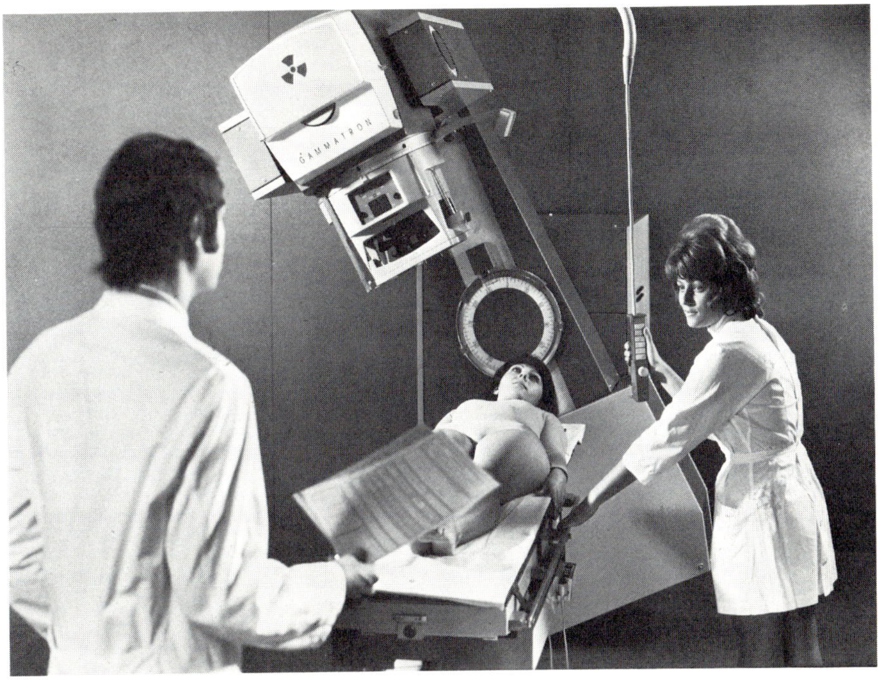

Figure 32.1
Radiocobalt therapy. (Courtesy Siemens Ltd.)

tained inside a lead shield. After careful alignment and withdrawal of the operator from the room, a beam of γ radiation is directed at the deep-seated growth. By rotating the radioactive source round the axis of the growth (Fig. 32.2) it is possible to give a higher dose of irradiation to the tumor than to the surrounding normal tissue.

Radioiodine is effective in treating that form of thyroid cancer which resembles a normal thyroid gland in its ability to take up iodine from the blood. It thus delivers its cell-destroying radiations exactly where they are needed.

Radiophosphorus, ^{32}P, tends to be localized in the blood-forming tissues of the bone-marrow and is consequently used in the treatment of polycythemia vera (overactivity of the marrow with excessive production of cells, particularly red blood cells).

Radiomimetics

Some chemicals, such as mustard gas (a chemical warfare agent) produce very similar biological effects to α, β, γ, and X rays.

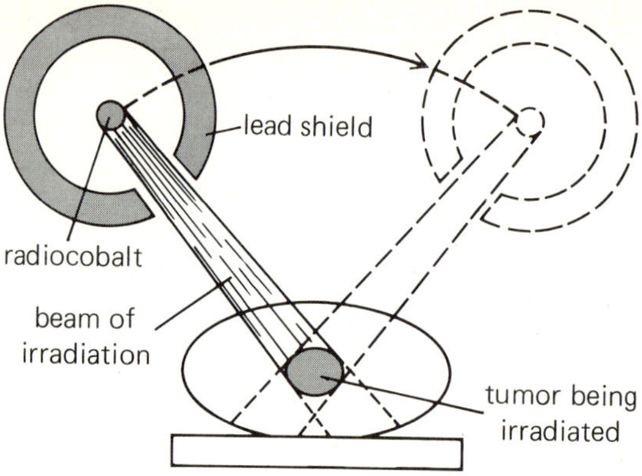

Figure 32.2
Irradiation of a deep-seated tumor with radiocobalt.

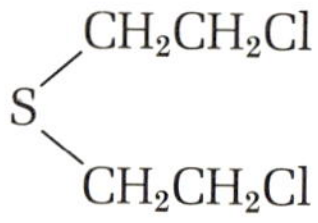

MUSTARD GAS

They are therefore known as radiomimetics, since they mimic the effects of these radiations. A few of these substances are used in the treatment of leukemia, but it is seldom possible to give sufficient doses to kill all the leukemic cells, as radiomimetics damage the intestine and bone marrow. They are not yet very suitable for the treatment of most tumors because their action is difficult to localize. However they offer considerable hope for the treatment of cancer by drugs in the future and have already provided a means of controlling leukemia for long periods. The search by chemists for a better drug usually entails modifying the molecular structure of the existing drug, preserving those atomic groupings which are known to be effective (in this case, the -CH_2CH_2Cl groups) until a molecule is built with the required properties and no unpleasant side-effects. The search for improved radiomimetics continues.

Future developments in radiotherapy

When deep-seated tumors are treated by conventional X rays, much of the energy is dispersed in the overlying normal tissues. This means that these normal tissues suffer radiation damage and that the tumor itself is likely to receive insufficient radiation to destroy it. If it were possible to use certain subatomic particles instead of X rays they would have the advantage that when travelling fast through the normal tissues they would release relatively little energy and also that as they slow down, for example, on reaching the tumor, their energy release would rapidly increase to reach a peak (the Bragg peak) at the moment of stopping.

Subatomic particles called pi-mesons not only show this Bragg peak effect but at the same time emit mesic X rays and finally cause miniature explosions as the pi-meson particles are converted into energy. A linear proton accelerator half a mile long is now being developed at Los Alamos, New Mexico, to produce such particles. Clinical trials have also been undertaken at Hammersmith, London, using neutrons to destroy human tissue. Although neutrons behave rather similarly to conventional X rays the initial results suggest that they may be more successful in preventing tumor recurrence than X rays. Such knowledge, improved radiomimetics and a greater understanding of the factors involved in the body's own immune processes for rejecting unwanted cells should enable us to control cancer more effectively.

Appendix

ELECTRONIC STRUCTURE OF THE ELEMENTS IN THEIR NORMAL STATES

ATOMIC NUMBER	ELEMENT	SYMBOL	NUMBER OF ELECTRONS IN SHELLS K	L	M	N	O	P	Q
1	Hydrogen	H	1						
2	Helium	He	2						
3	Lithium	Li	2	1					
4	Beryllium	Be	2	2					
5	Boron	B	2	3					
6	Carbon	C	2	4					
7	Nitrogen	N	2	5					
8	Oxygen	O	2	6					
9	Fluorine	F	2	7					
10	Neon	Ne	2	8					
11	Sodium	Na	2	8	1				
12	Magnesium	Mg	2	8	2				
13	Aluminum	Al	2	8	3				
14	Silicon	Si	2	8	4				
15	Phosphorus	P	2	8	5				
16	Sulfur	S	2	8	6				

ELECTRONIC STRUCTURE OF THE ELEMENTS *(continued)*

ATOMIC NUMBER	ELEMENT	SYMBOL	NUMBER OF ELECTRONS IN SHELLS						
			K	L	M	N	O	P	Q
17	Chlorine	Cl	2	8	7				
18	Argon	Ar	2	8	8				
19	Potassium	K	2	8	8	1			
20	Calcium	Ca	2	8	8	2			
21	Scandium	Sc	2	8	9	2			
22	Titanium	Ti	2	8	10	2			
23	Vanadium	V	2	8	11	2			
24	Chromium	Cr	2	8	13	1			
25	Manganese	Mn	2	8	13	2			
26	Iron	Fe	2	8	14	2			
27	Cobalt	Co	2	8	15	2			
28	Nickel	Ni	2	8	16	2			
29	Copper	Cu	2	8	18	1			
30	Zinc	Zn	2	8	18	2			
31	Gallium	Ga	2	8	18	3			
32	Germanium	Ge	2	8	18	4			
33	Arsenic	As	2	8	18	5			
34	Selenium	Se	2	8	18	6			
35	Bromine	Br	2	8	18	7			
36	Krypton	Kr	2	8	18	8			
37	Rubidium	Rb	2	8	18	8	1		
38	Strontium	Sr	2	8	18	8	2		
39	Yttrium	Y	2	8	18	9	2		
40	Zirconium	Zr	2	8	18	10	2		
41	Niobium	Nb	2	8	18	12	1		
42	Molybdenum	Mo	2	8	18	13	1		
43	Technetium	Tc	2	8	18	14	1		
44	Ruthenium	Ru	2	8	18	15	1		
45	Rhodium	Rh	2	8	18	16	1		
46	Palladium	Pd	2	8	18	18			
47	Silver	Ag	2	8	18	18	1		
48	Cadium	Cd	2	8	18	18	2		
49	Indium	In	2	8	18	18	3		
50	Tin	Sn	2	8	18	18	4		
51	Antimony	Sb	2	8	18	18	5		
52	Tellurium	Te	2	8	18	18	6		
53	Iodine	I	2	8	18	18	7		
54	Xenon	Xe	2	8	18	18	8		
55	Caesium	Cs	2	8	18	18	8	1	
56	Barium	Ba	2	8	18	18	8	2	
57	Lanthanum	La	2	8	18	18	9	2	
58	Cerium	Ce	2	8	18	19	9	2	
59	Praseodymium	Pr	2	8	18	21	8	2	
60	Neodymium	Nd	2	8	18	22	8	2	

ELECTRONIC STRUCTURE OF THE ELEMENTS *(continued)*

ATOMIC NUMBER	ELEMENT	SYMBOL	NUMBER OF ELECTRONS IN SHELLS K	L	M	N	O	P	Q
61	Promethium	Pm	2	8	18	23	8	2	
62	Samarium	Sm	2	8	18	24	8	2	
63	Europium	Eu	2	8	18	25	8	2	
64	Gadolinium	Gd	2	8	18	25	9	2	
65	Terbium	Tb	2	8	18	26	9	2	
66	Dysprosium	Dy	2	8	18	28	8	2	
67	Holmium	Ho	2	8	18	29	8	2	
68	Erbium	Er	2	8	18	30	8	2	
69	Thulium	Tm	2	8	18	31	8	2	
70	Ytterbium	Yb	2	8	18	32	8	2	
71	Lutetium	Lu	2	8	18	32	9	2	
72	Hafnium	Hf	2	8	18	32	10	2	
73	Tantalum	Ta	2	8	18	32	11	2	
74	Tungsten	W	2	8	18	32	12	2	
75	Rhenium	Re	2	8	18	32	13	2	
76	Osmium	Os	2	8	18	32	14	2	
77	Iridium	Ir	2	8	18	32	15	2	
78	Platinum	Pt	2	8	18	32	17	1	
79	Gold	Au	2	8	18	32	18	1	
80	Mercury	Hg	2	8	18	32	18	2	
81	Thallium	Tl	2	8	18	32	18	3	
82	Lead	Pb	2	8	18	32	18	4	
83	Bismuth	Bi	2	8	18	32	18	5	
84	Polonium	Po	2	8	18	32	18	6	
85	Astatine	At	2	8	18	32	18	7	
86	Radon	Rn	2	8	18	32	18	8	
87	Francium	Fr	2	8	18	32	18	8	1
88	Radium	Ra	2	8	18	32	18	8	2
89	Actinium	Ac	2	8	18	32	18	9	2
90	Thorium	Th	2	8	18	32	18	10	2
91	Protoactinium	Pa	2	8	18	32	20	9	2
92	Uranium	U	2	8	18	32	21	9	2
93	Neptunium	Np	2	8	18	32	22	9	2
94	Plutonium	Pu	2	8	18	32	23	9	2
95	Americium	Am	2	8	18	32	24	9	2
96	Curium	Cm	2	8	18	32	25	9	2
97	Berkelium	Bk	2	8	18	32	26	9	2
98	Californium	Cf	2	8	18	32	27	9	2
99	Einsteinium	Es	2	8	18	32	28	9	2
100	Fermium	Fm	2	8	18	32	29	9	2
101	Mendelevium	Md	2	8	18	32	30	9	2
102	Nobelium	No	2	8	18	32	31	9	2
103	Lawrencium	Lw	2	8	18	32	32	9	2

Index

76 77 78 79 9 8 7 6 5 4 3 2 1